NATIONAL
GEOGRAPHIC

POCKET GUIDE TO THE
Mammals
OF NORTH AMERICA

NATIONAL GEOGRAPHIC

POCKET GUIDE TO THE
Mammals
OF NORTH AMERICA

CATHERINE HERBERT HOWELL

NATIONAL GEOGRAPHIC
WASHINGTON, D.C.

CONTENTS

||||||||||||||||||||||||||||||||||||||||||||||||||||||||||||||||||||||||||||||||||||||

Mammals
The Summit of Diversity

Our closest mammalian relatives, the nonhuman primates, are not native to the United States and Canada, but kinship pervades the anatomy, physiology, and behaviors of the many mammal species that are. We understand the fierce protectiveness of a mother bear toward her cubs, the driving urge of a Moose to find a mate, and a small rodent's need to "make hay while the sun shines" ahead of a long winter.

What Is a Mammal?

Mammals inhabit a wide range of shapes and sizes, from a 170-ton (154-tonne) whale to a bat the size of a bumblebee, but they all share certain characteristics. Mammals form a large class of warm-blooded vertebrates that have hair of some sort at some time of life (even whales and dolphins), three middle-ear bones, a neocortex in the brain, and mammary glands in the females that produce milk to nourish young. Most mammals bear young sustained by a placenta during pregnancy. Monotremes, an ancient order, are one exception: Their offspring develop in the leathery eggs they lay that, upon hatching, are nursed from pores on the mother's belly. These unique mammals—the Duck-billed Platypus and several species of echidna, or "spiny anteater"—are absent from the New World. Another exception is the marsupial, which gives birth to underdeveloped young that finish their development in a maternal pouch where nursing takes place. The Virginia Opossum is the only marsupial native to our area.

Placental mammals produce two kinds of young: precocial and altricial. Precocial young are basically miniatures of their parents—furred and functional and ready to move with adults on only several hours' or days' notice. Altricial young enter the world naked, often pink and wrinkled, with eyes closed shut. They need time to reach adulthood; that can be a matter of weeks in some rodents, or a matter of many years in primates.

Shy, solitary creatures by nature, Raccoons spend their days high in the trees and forage by night.

Mammals in Your World

North America north of Mexico, the scope of this guide, enjoys a wealth of mammalian species, some 474 representing 12 different orders, or related groups of mammals. They are scattered across the continent, inhabiting every ecological niche from Arctic tundra to subtropical swamp. Depending on where you live, you might see a few or a lot of mammals on a daily basis. Most people probably see a lot of only a few species. Ubiquitous mammals, such as tree and ground squirrels, are part of the background of urban and rural life in many parts of the United States and Canada. Other species are seen on occasion—the Raccoon we startle when we take out the trash, or the fox glimpsed in an arcing pounce in a grassy field. But looking for and recognizing the mammals that we don't normally see requires guidance.

Species at the larger end of the mammalian scale are easier to identify: A Grizzly Bear, a Mountain Goat, and a bull Moose are virtually unmistakable. At the smaller end, things become more difficult. Among the rodents, some species are so similar—in the genus Peromyscus, the deermice, for example—that you'd need to see their dental records (or intact skull specimens) to tell them apart. Even scientists find this very challenging.

Identifying Mammals

This guide will help you learn about mammals and how to identify them, as well as provide the potential to observe them based on their range, habitats, and habits. The species accounts in this guide introduce 160 mammal species found in North America north of Mexico. It is a small but representative sample of the total number of species. These accounts include species from all the major mammalian groups except those that live their lives largely or exclusively in the water, such as whales, seals, manatees, and sea otters.

The sections present the species in groups arranged according to a current understanding of taxonomic classification, an understanding that is under constant revision due to the advances of molecular biology and that is not without its controversies. Scientific names provided generally follow Wilson and Reeder's Mammal Species of the World, 3rd edition (2005), and the common names that go with them are those with wide recognition. Within each genus presented, the species are listed in alphabetical order of their names, as not all species have been investigated equally and their interrelationships are imperfectly known. The measurements include the length of an animal's head, body, and tail. Height at shoulder is the standard measurement shown for the hoofed mammals. Each chapter also offers a set of characteristics that will help you identify a species and compare it with others. The species accounts also provide details about the animal's habitat, range, and diet.

On the Right Track

Many mammals lead lives largely hidden from our view—the vast majority are nocturnal or crepuscular (active at dawn and dusk)—but we know they are there by the signs that they leave. Signs can range from their tracks, to their droppings (known as scat), to their communication methods, to leftovers from meals, to engineering projects such as a beaver dam. All of these activities leave clues that become easy to spot if you know where and how to look.

Tracks point us in the direction of a mammal's presence, whether they are deposited

Distinct trail patterns help trackers to a glimpse of the Red Fox's quiet life.

in mud, snow, or as wet marks left on a concrete sidewalk. Learning about mammalian appendages, how mammals walk, run, or hop, and the patterns they make as they move aid identification. Each account in this guide includes the tracks made by the species; wing shape is shown for bats.

Whole books have been written on scat. Different groups of mammals have different sizes and shapes of scat with different contents. The small, regular pellets of rabbits and hares bear little resemblance to the large, lumpy plops of a Moose. Scat of the same species can show seasonal differences: the Red Fox's looser, berry-rich summer droppings look very different from the fur-matted rope of its winter scat. How scat is deposited also provides clues. Members of the cat family, for example, usually scrape the substrate before defecating, so you may find their scat in an excavated dent. Foxes often defecate in the middle of a trail or at a trail crossing. Among scat detectives, where and how is as important as what.

Many mammals leave calling cards in the form of gnawing marks and scrapes on tree trunks. These suggest feeding behaviors, construction techniques, or communication methods used to demarcate territory. Other signs to be aware of include small, flattened areas in long grass that could be the forms, or nest depressions, of rabbits, or shallow, oval pits of many sizes in dust or sand that could be evidence of wallows or dust baths of species from bison to mice. Tracks and signs form a whole category of wildlife knowledge, and many excellent books can provide a comprehensive overview.

Mammals in Perspective

The mammalian landscape of North America has changed greatly from the last ice age, when mammals such as mammoths, mastodons, and saber-toothed cats roamed much of the continent. Theories of the demise of these megafauna implicate climate change, habitat change, pathogens, and human impact. Closer to our time, human enterprise nearly eliminated one of the continent's most spectacular mammals, the American Bison. In the 19th century, these magnificent creatures still populated large swaths of the Great Plains, hunted sustainably for the most part by native peoples for centuries. But the buffalo got in the way of westward expansion by human newcomers, and financial gain was a strong incentive to hunt them extensively. In only 40 years, tens of millions were reduced to a thousand.

It is still possible to see great assemblages of mammals on our continent. In the Arctic, hundreds of thousands of caribou travel together for hundreds of miles to seasonal feeding grounds. Along the way, they pass under portions of oil pipeline elevated for their accommodation. At dusk in the summer months, a seemingly endless cloud of more than a million free-tailed bats emerges from under a bridge in Austin, Texas, to begin their nightly assault on swarms of flying insects.

As climate change and habitat disruption cause mammals to move into areas new to them, some species are appearing in places you might not expect, such as the Coyote in the environs of the United States capital. That icon of the Southwest may have followed burgeoning populations of deer and rabbits as it drifted closer to Washington, D.C. And while many people warm to the idea of more wildlife in the cities, its presence there indicates that we have failed to maintain natural habitats. Not all migrations are successful for the migrants or the native and long-settled species—whether animals or plants—as the perspective of time has made abundantly clear.

MAMMALS IN MOTION

Mammalian limbs display a wide range of adaptations for locomotion and other activities. Bat wings, seal flippers, and beaver webbed hind feet allow mastery of air or water. Strong, articulate feet aid tree climbing and object manipulation in opossums and raccoons. Long hind feet help hares and kangaroo rats escape predators in elusive bounds. Prairie dogs and moles use limbs to excavate homes and tunnels. Bears amble on large paws equipped with lethal claws, whereas Mountain Lions keep their feline claws retracted while moving. Horses and other hoofed mammals need stable support for all-terrain travel and long hours of grazing and browsing. Mountain Goat hooves act like suction cups on rocky heights.

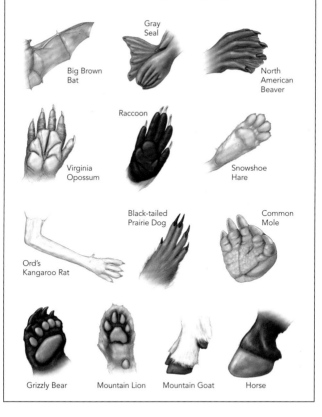

Big Brown Bat

Gray Seal

North American Beaver

Virginia Opossum

Raccoon

Snowshoe Hare

Ord's Kangaroo Rat

Black-tailed Prairie Dog

Common Mole

Grizzly Bear

Mountain Lion

Mountain Goat

Horse

Virginia Opossum

Didelphis virginiana L 24–30 in (61–76 cm)

The only marsupial native to the United States and Canada, the Virginia Opossum lacks the cuddle factor of its Koala cousin, but it adapts well to its increasingly suburban and urban lifestyle.

KEY FACTS

Body is grayish, with long guard hairs; nose and clawed toes pink; naked prehensile tail.

+ habitat: Forests, old fields, suburban and urban areas

+ range: Central and eastern U.S. to southeastern Canada; introduced elsewhere

+ food: Grasses, nuts, fruits, insects, snakes, and carrion

The opossum's famous play-dead routine is very convincing: A motionless body, vacant stare, and lolling tongue fool predators such as foxes, Bobcats, and Coyotes. This nocturnal species climbs agilely using the opposable thumbs of its hind feet, but waddles clumsily along the ground. It inhabits holes, crevices, and other animal dens. Twice a year an adult female may give birth to up to 20 bee-size, blind, naked newborns that crawl to her fur-lined pouch. Mortality is high. The young nurse in the pouch for about three months and ride on the mother's back when outside. Opossums do not hibernate, but become less active in winter.

Nine-banded Armadillo/
Common Long-nosed Armadillo
Dasyapus novemcintus L 25–32 in (64–81 cm)

An exclusive member of living mammals with bony shells, the Nine-banded Armadillo is the only armadillo in the United States.

KEY FACTS

Scaly body has 8 to 9 overlapping, movable bands on back; underside is furred; tail is armored.

+ habitat: Fields, woodlands, brushy areas, and roadsides

+ range: Southeastern United States; range is expanding

+ food: Insects, spiders, other invertebrates, small vertebrates, and carrion

Though it looks encumbered by its unique, scaly armor, the armadillo is flexible—and adaptable. It runs and dodges with a stiff-legged gait and can burrow quickly to escape. It can hold its breath for six minutes to cross a riverbed, weighted down by its shell, or can inflate by swallowing air to float across. Within a burrow or an aboveground nest, a female bears one litter a year of four identical offspring. Armadillo eyes don't shine in the dark, probably increasing the species' mortality rate along southern roads. Armadillos also jump straight up when startled, causing crushed backs when a car straddles a road-crossing armadillo.

Sewellel/Mountain Beaver
Aplodontia rufa L 14 in (36 cm)

Previously called the Mountain Beaver, this primitive rodent is
not a beaver, but is the only member of its genus and family,
the Aplodontidae. Its lineage extends 50 million years.

KEY FACTS

Body above is dark
brown or blackish;
underside is lighter;
white spot at base
of ear; very short tail.

+ **habitat:** Thickets,
meadows, and forests
near water

+ **range:** Southern
British Columbia to
California

+ **food:** Needles and
twigs of conifers and
herbaceous plants,
including ferns

The elusive and mainly nocturnal Sewellel looks like a
Muskrat with a stubby tail. It digs tunnels that can be
several hundred feet long with multiple entrances. Piles of
dirt often mark the entrances, which frequently are close
to water. The Sewellel also climbs trees in search of tasty
twigs and needles, ascending 20 feet (6 m) up Lodgepole
Pines and descending by using stubs of
cut branches as ladder rungs. Sewellel
females are late breeders; they have
a first litter of between one and six
young at about two years old. The
common name likely comes from a
name for fur robes that Native Ameri-
cans of the Pacific Northwest wore.

Tassel-eared Squirrel

Sciurus aberti L 18 in (46 cm)

The subspecies of Tassel-eared Squirrel known as Abert's Squirrel lives south of the Grand Canyon, and the Kaibab subspecies lives north of it. Brown and black melanistic forms also occur.

KEY FACTS

Body is gray, with reddish sides and white (Abert's) or black (Kaibab) underside.

+ **habitat:** Ponderosa Pine and other coniferous or mixed forests

+ **range:** Southwestern Wyoming, Colorado, New Mexico, Arizona, and Utah

+ **food:** Pine nuts, buds, bark, berries, fungi, bones, and carrion

The Tassel-eared Squirrel is known for the long ear tufts that are especially noticeable in winter. It also sports a lush, shortish tail that is gray on top and white elsewhere in the Abert's and all white in the Kaibab subspecies. This diurnal squirrel has a close association with the Ponderosa Pine, husking pinecones for seeds and eating the inner bark of twigs that it cuts. A pile of denuded twigs under a pine is a classic sign of its presence. It builds twig nests in trees where the adult female raises broods of two to four young. Mating is preceded by a chase involving a dominant male and his subordinates.

Eastern Gray Squirrel

Sciurus carolinensis L 18.5–26.5 in (47–67 cm)

The Eastern Gray Squirrel swishes its versatile bushy tail in courtship, flicks it in greeting, fluffs it in anger, uses it as an umbrella when it rains, and curls it around its body in the cold.

KEY FACTS

Body is gray or brownish; tail is edged in white; melanistic animals are dark brown or black.

+ **habitat:** Hardwood and mixed forests, suburban and urban areas

+ **range:** Eastern U.S. to southern Canada; introduced in West

+ **food:** Acorns, other nuts, seeds, buds, fruit, fungi, insects, and birds

Eastern Grays build leafy nests in trees, known as dreys, which are visible among bare branches in winter. Some are basic sleeping platforms; others are domed and secured for winter living. This diurnal species usually lives solitarily, but individuals may bed down together in winter while remaining active in the daytime. They breed twice a year; courtship involves spirited chasing, and a female may mate with more than one male. Litters average two to four young. Buried acorns are retrieved randomly by their odor. The town of Olney, Illinois, harbors a long-established, protected population of albino, or white, *S. carolinensis*.

Western Gray Squirrel

Sciurus griseus L 22 in (56 cm)

Western Gray Squirrels look a lot like their eastern counterparts, but are bigger and grayer. They're also shyer and more secretive, and haven't taken in a big way to urban life.

KEY FACTS

Body is gray; eyes are ringed in white; underside is white; bushy dark gray tail is edged in white.

+ habitat: Mainly oak, coniferous, and mixed forests

+ range: Southern California to northern Washington

+ food: Acorns, pine nuts, berries, vegetation, fungi, bark, insects, birds, and eggs

These squirrels have a preference for pine nuts. They clip cones from trees and strip them of their seeds on the ground. Western Grays may rest together in a twig nest on a tree branch, known as a drey, but otherwise adults mingle only during mating time. From two to six young are often born in a tree cavity, but heavy flea infestation can cause the mother to move them to a drey. The species sometimes raids almond and walnut orchards, with economic consequences to growers. It competes for food and habitat with the more adaptable and gregarious Eastern Gray Squirrel, which has been introduced to parts of the range.

Eastern Fox Squirrel

Sciurus niger L 20–24 in (51–61 cm)

Taxonomist Linnaeus first dubbed the Fox Squirrel *niger,* believing that the black form represented the species. The most widespread form is largely orangish brown.

KEY FACTS

Colors are variable, often yellowish brown with orange cheeks, eye rings, and underside.

+ habitat: Open woodlands, golf courses, and suburbs

+ range: Eastern and central U.S., southern Canada; introduced elsewhere

+ food: Acorns, other nuts, seeds, buds, roots, fruits, bulbs, insects, and eggs

The largest squirrel species in the East, the Eastern Fox Squirrel weighs up to 3 pounds (1.4 kg), and it has a bushy tail that is as long as its body. Although it primarily caches and eats acorns, it is an eclectic feeder, relishing maple sap and corn on the cob. Populations have been severely reduced in the Northeast due to hunting and forest clearing. The Delmarva Fox Squirrel, a gray subspecies, is designated as endangered. Melanistic, or black, populations are found mainly in the South. The Eastern Fox Squirrel in general has extended its range and been introduced successfully in the West.

Douglas Squirrel

Tamiasciurus douglasii L 12 in (30 cm)

A resident of Pacific coast forests, the small Douglas Squirrel shares a bit of range and a number of habits, including general noisiness, with the American Red Squirrel.

KEY FACTS

Back is brownish or reddish gray; a black stripe separates it from white, grayish, or orangish underside. The ears have black tufts in winter.

+ habitat: Coniferous and mixed forests

+ range: Southern California to British Columbia

+ food: Seeds, fungi, flowers, insects, young birds, eggs, and bones

The Douglas Squirrel is known for its wide array of vocal stylings. Naturalist John Muir called it the "mockingbird of squirrels . . . barking like a dog, screaming like a hawk, chirping like a blackbird or a sparrow." The species is a cone clipper, harvesting cones from species such as fir, pine, spruce, and hemlock. Most cones are cached in a midden hoard for winter. Named for Scottish botanist David Douglas, namesake of the fir, this species nests in tree holes lined with shredded bark. A litter of four to six is born in early summer; females may have an additional litter in the fall, depending on location.

American Red Squirrel

Tamiasciurus hudsonicus L 12 in (30 cm)

The American Red Squirrel chatters loudly as a matter of routine and will scold intruders into its territory. Ojibwe Indians called the species *ajidamo*, roughly meaning "tail in air."

KEY FACTS

Body is russet above; underside is white; white eye rings; grizzled orange tail.

+ habitat: Coniferous and mixed forests, parkland, and hedgerows

+ range: Alaska and Canada into northeastern U.S., Rockies, and Appalachians

+ food: Pine nuts, berries, fungi, insects, birds, eggs, and sap

The American Red Squirrel is much less social than the Eastern Gray Squirrel and is not very tolerant of others of its own species. Its feeding habits leave a telltale sign of its presence. It clips cones from conifers and stores them to ripen underground, under rocks, or in holes until it is time to take them to a feeding station on a branch and systematically strip the seeds. The dropped scales form a midden underneath, up to 3 feet (0.9 m) high and perhaps 30 feet (9 m) long. Once a year—or twice, depending on location—males enter a breeding female's territory to mate. She will bear a litter of three to five young.

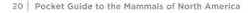

||||||||||||||||||||||||||||||||||||||||||||||||||||||||||||||||||||||||||||||

Northern Flying Squirrel

Glaucomys sabrinus L 12 in (30 cm)

Flying Squirrel is a misnomer, because it doesn't fly but glides from tree to tree by stretching out a membrane between front and hind feet. The species can cover up to 300 feet (90 m) in a glide.

KEY FACTS

Body is brownish; membrane edge is black; tail is flat. A Pacific Northwest subspecies is darker.

+ habitat: Coniferous and mixed woodlands

+ range: Alaska through Canada into northern U.S., and southward in western and eastern mountains

+ food: Nuts, seeds, sap, fungi, lichens, birds, and eggs

Larger and heavier than its southern counterpart, the Northern Flying Squirrel shares some parts of its range with the Southern Flying Squirrel. Like the Southern, the Northern is nocturnal, coming out in the evening to partake largely of the offerings found on a tree, sometimes including birds' eggs and nestlings. This species is generally quieter in voice and less anxious in its actions than the southern one. It nests in tree cavities and builds its own nest, using twigs and leaves, lining it with finer materials, and sometimes enclosing a bird's nest. A female has one litter of two to four young in late spring.

Southern Flying Squirrel

Glaucomys volans L 9 in (23 cm)

The Southern Flying Squirrel peers from a tree hole, planning a route. It glides with membranes stretched, shifting its body to steer, then pulls up for a four-point landing on its destination tree.

KEY FACTS

Body is pale grayish brown; cheeks and underside white; membrane edge black.

+ habitat: Oak-hickory and mixed woodlands, suburban and urban areas

+ range: Eastern half of U.S. into southern Canada

+ food: Nuts, seeds, flowers, berries, bark, fungi, lichens, insects, birds, and birds' eggs

The Southern Flying Squirrel is smaller and paler than the Northern Flying Squirrel, but the two species have similar habits. The furred membrane between front and rear legs gives this species some 50 square inches (323 sq cm) with which to glide. The membrane hangs in accordion folds when not in use. Flying squirrels nest in tree cavities and use nest boxes placed in trees, which they will line with materials such as fur and feathers. Females have litters of three to five young after a 40-day gestation; those in the South may have two litters a year. These social squirrels may nest in large groups in winter.

Harris's Antelope Squirrel

Ammospermophilus harrisii L 9.5 in (24 cm)

Antelope Squirrels look somewhat like chipmunks, but lack the face stripes. The genus name of these desert species, *Ammospermophilus,* means "lover of sand and seeds."

KEY FACTS

Body is grayish brown above, with a white stripe on the sides; eyes are circled in white; tail is uniformly grizzled.

+ habitat: Desert areas with cacti and desert shrubs

+ range: Arizona and southwestern New Mexico

+ food: Cactus seeds and fruit, yucca seeds, mesquite beans, and insects

Harris's Antelope Squirrel can easily navigate prickly cacti of its desert home. It can climb to the top of a Cholla cactus without getting spines in the soft pads of its feet. It may even sit on the Cholla's summit to get a good look around. The species is diurnal, even in the heat, alternating bursts of activity with rest periods. Its burrow is frequently found at the base of a desert bush. It does not hibernate, but remains active year-round, and its body weight does not fluctuate seasonally. Breeding season may start as early as December; a litter of five to nine young is born after a one-month gestation.

White-tailed Antelope Squirrel

Ammospermophilus leucurus L 9 in (23 cm)

The White-tailed Antelope Squirrel folds its two-tone tail over its back to create shade, one of the adaptive strategies this desert species displays.

KEY FACTS

Back is grayish; side has narrow white stripe; tail is grizzled on top.

+ habitat: Desert shrubland, sandy and rocky areas

+ range: Southern Oregon and Idaho to southern California, northern Arizona, and northwestern New Mexico

+ food: Vegetation, seeds, insects, and other invertebrates

White-tailed Antelope Squirrels have a number of physiological adaptations to their dry desert environment, including several that conserve moisture, as they depend on drinking water. Active during the day, they sprint from one shady area to another to forage. Individual White-tails use a number of different burrows, often the abandoned burrows of kangaroo rats. This species does not hibernate, but grows a warm coat in cold areas and may refrain from activity during bad weather. A group may huddle in a burrow to stay warm. White-tails observe a dominance hierarchy rather than squabble over territories.

White-tailed Prairie Dog

Cynomys leucurus L 14 in (36 cm)

This mountain species leads a much less social life than the Black-tailed Prairie Dog. It lives in smaller colonies, and it ranges as high as 12,000 feet (3,700 m).

KEY FACTS

Back is grayish yellow; sides are yellowish; underside is buff-orange; tail has a white tip.

+ **habitat:** Grasslands and shrublands

+ **range:** Southern Montana, Wyoming, northeastern Utah, and northwestern Colorado

+ **food:** Grasses, sedges, other herbaceous plants, and woody plants

The White-tailed Prairie Dog builds complex underground chambers in its mountain habitat. This species does not require extensive clearings as the more vulnerable plains prairie dogs do for keeping a vigilant eye on the landscape. These animals spend the summer feeding voraciously, preparing for the long sleep of hibernation. By August or September many of the adults have bedded down underground; juveniles may linger above until heavy snows arrive. Four to six pups are born in May. White-tails have been targeted for extermination and fall prey to eagles, hawks, Bobcats, and badgers. They also are susceptible to plague.

Black-tailed Prairie Dog

Cynomys ludovicianus L 16 in (41 cm)

Millions of Black-tailed Prairie Dogs once excavated towns from southern Canada to the Rio Grande. Reduced in numbers, they have the most socially complex ground squirrel society.

KEY FACTS

The back is orangish brown; the sides and belly are paler; the narrow tail is tipped in black.

+ **habitat:** Short-grass prairies and other grasslands

+ **range:** Southern Saskatchewan to southwestern Texas

+ **food:** Roots and vegetation of grasses and other herbaceous plants

Prairie dog towns are vast constructions consisting of tunnels, specialized chambers, and multiple entrances, which serve as lookout points. Towns are divided into sectors containing one adult male and associated females and young; these groups hold off encroaching neighbors. Prairie dogs communicate with barks, chuckles, and a long *weee-oooo* whistle emitted during a "jump-yip," the species' signature move that may say "all's clear." Ranchers and farmers tended to despise prairie dogs for eating grasses and making town entrance holes that hobbled horses. Targeted elimination and habitat loss greatly reduced their numbers.

Hoary Marmot
Marmota caligata L 30 in (76 cm)

French-Canadian trappers called the Hoary Marmot *le siffleur*—
the whistler—for the shrill warning notes that resonate in mountain
meadows. Its vocal repertoire also includes barks and hisses.

KEY FACTS

Front half of body is
whitish, rear half is
yellowish or reddish
brown; black on snout
and crown.

+ habitat: Mountain
meadows, rocky areas,
and talus slopes

+ range: Alaska and
northwestern Canada
into Washington,
Idaho, and Montana

+ food: Grasses,
sedges, and other
herbaceous plants

Our largest marmot species, the Hoary Marmot prefers
to make its burrow on a warm, south-facing slope,
often under a boulder. The chunky animal may hibernate
up to nine months of the year and emerge in late spring
ready to mate. Adult males take on and defend several
females, and they all live in a group with their offspring.
In the tolerant Hoary Marmot family,
the young are not expelled as they
are in many other rodent species,
and may remain with the group
for two years. Females breed every
other year, giving birth to two to four
young. The species name, *caligata*, refers to
the black boots Roman soldiers wore.

Yellow-bellied Marmot/Rockchuck

Marmota flaviventris L 23 in (58 cm)

Despite the name Yellow-bellied, this marmot is anything but timid. It is noted for its raucous vocalizations, including whistles, screams, and tooth chattering.

KEY FACTS

Back is grayish brown; underside and lower legs are yellowish to bright orange; tail is reddish to dark brown; sides of neck have buff patches.

+ habitat: Mountain meadows in rocky areas

+ range: Western U.S. into southern Canada

+ food: Mainly flowers, grasses, other plants, and seeds

The Yellow-bellied Marmot ranges to 12,000 feet (3,700 m). Home is often a burrow in a rocky area with a nearby lookout boulder. Males often take multiple mates—one had a record of 31— and live with them and their offspring, born in litters of three to eight about a month after mating. The marmots spend the summer filling up in alpine meadows before hibernating for as long as nine months in cold climates. Those in warmer areas may estivate, or spend time in torpor, during hot summers. Some marmots at national and state parks have turned into panhandlers, sitting on their ample backsides on rock perches and playing to the tourists.

Groundhog/Woodchuck
Marmota monax L 24 in (61 cm)

Punxsutawney Phil is Pennsylvania's famous weather prognosticator. Many Groudhogs, aka Woodchucks, will hibernate in burrows— if left undisturbed—well past February 2, Groundhog Day.

KEY FACTS

Back is grizzled gray-ish or brownish; nose, eyes, and ears are located on the same plane.

+ **habitat:** Fields, woodland edges, stream banks, and roadsides

+ **range:** Eastern U.S. and Canada through central Canada into Alaska

+ **food:** Grasses and forbs, insects, and crops such as clover and alfalfa

The Groundhog, a kind of ground squirrel known as a marmot, is a solitary creature. Sometimes an early morning or late afternoon drive down a highway bounded by a grassy verge will reveal single Groundhogs at almost regular intervals, foraging among the many kinds of plants they enjoy. They eat heartily in summer to put on a layer of fat that will maintain them through hibernation, a remarkable physi-ological adaptation. The body temperature of a hibernating Wood-chuck may dip to 35°F (2°C), and its heart may slow to four beats a minute. Woodchucks are champion tunnel diggers and can burrow out of sight in a minute.

Columbian Ground Squirrel

Spermophilus columbianus L 13 in (33 cm)

Both males and females of the highly social Columbian Ground
Squirrel defend real estate within their home ranges. To secure
their young, nesting females plug up their nest chambers at night.

KEY FACTS

Back is grizzled
brownish; underside,
legs, and top of snout
are orange; eye rings
are white.

+ habitat: Wet mead-
ows and grasslands,
clear-cut areas, and
rangeland

+ range: Mountains
of northwestern U.S.
into British Columbia
and Alberta

+ food: Flowers,
seeds, bulbs, fruit,
and vegetation

Group members in this very territorial species greet
each other by "kissing" when they meet. They also
scent-mark liberally, using multiple scent glands on the
head and body to mark their turf. These animals cram all
activity into about a hundred days before heading underground to hibernate. A great deal of care goes into preparing the hibernation den. Usually
it is a carefully lined and domed
chamber that incorporates a drain
to help keep it dry. Young tend to
hibernate with their mothers. Adult
males rely on a store of food when
they emerge ahead of the females,
if food aboveground is scarce.

Franklin's Ground Squirrel

Spermophilus franklinii L 14 in (36 cm)

At a glance, this species resembles the Eastern Gray Squirrel, but look closely and you'll see that the ears are much smaller and the tail is thinner and less bushy.

KEY FACTS

Back is grizzled grayish brown or yellowish; underside is grayish brown or white; tail is narrow and bushy.

+ habitat: Grasslands, thickets, woodland and marsh edges, and embankments

+ range: North-central U.S. to south-central Canada

+ food: Vegetation, seeds, insects, birds, eggs, toads, and carrion

Franklin's Ground Squirrel keeps a low profile in the tall grasses and dense vegetation it favors. It digs extensive burrows and tends to vanish underground at the first sign of trouble. These squirrels live solitarily or in loose aggregations. Franklin's Ground Squirrel partakes of a more eclectic—and carnivorous—diet than many other ground squirrels. The diet includes duck eggs from nests in prairie potholes. Much of the species' preferred habitat of tall-grass prairie has disappeared due to agriculture and urbanization, and it is listed as threatened or endangered in a number of states within its range.

Arctic Ground Squirrel
Spermophilus parryii L 14 in (36 cm)

Orange fur is no liability for this squirrel. Hibernating in the snowy
months, it needs no camouflage. Different calls signal a warning
about aerial or terrestrial predators.

KEY FACTS

The back is grizzled
reddish brown; the
snout, underside, and
limbs are a vibrant
orange-brown.

+ **habitat:** Tundra,
alpine meadows,
and sandy hills

+ **range:** Alaska and
northwestern Canada

+ **food:** Seeds, ber-
ries, grasses, leaves,
fungi, insects, and
carrion

Location is everything for the Arctic Ground Squirrel in
search of a construction site. It can build its labyrinthine burrows only where at least 3 feet (0.9 m) of unfro-
zen surface sit above the permafrost. It crafts specialized
burrows for different purposes, whether resting, hibernating, or raising young. A nesting burrow will be lined with
soft grasses and perhaps caribou
fur. This ground squirrel fattens
up in summer—and gnaws fallen
caribou antlers for minerals—
before it beds down for about six
months. It awakens in April or May
and may have to burrow through several feet
of snow to look for a mate.

California Ground Squirrel

Otospermophilus beecheyi L 16 in (41 cm)

Some California Ground Squirrels lead solitary lives in underground burrows. Others may live in connected colonies of up to a dozen adults, accessed by separate entrances.

KEY FACTS

Back is orangish brown grizzled with cream; white mantle on shoulders; underside and legs white to orangish.

+ habitat: Fields, pastures, cropland, semiarid areas

+ range: Southern Washington and western Oregon to California

+ food: Nuts, seeds, fruit, fungi, birds, insects, and garden vegetables

The California Ground Squirrel may spend up to two-thirds of the year underground, hibernating in winter and estivating in the summer. It emerges in the spring, enlarging its burrow and refurbishing it with fresh grasses. Males and females come together briefly to mate, and after a month, litters of three to nine young are born, which may remain active through the year. The species often feeds on convenient agricultural crops and vegetable plots, earning growers' wrath and making it a target for elimination. It also avails itself of roadside garbage bins. This mammal is a vector for plague and tularemia.

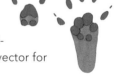

Rock Squirrel

Otospermophilus variegates L 20 in (51 cm)

A bushy tail helps to give this species the appearance of a tree squirrel, but the Rock Squirrel usually lives underground. It sometimes climbs trees, though, and occasionally nests in them.

KEY FACTS

Its colors vary, often grizzled brown; the underside is whitish; the tail is long and bushy.

+ **habitat:** Canyons, cliffs, and other rocky areas; bridges and stone walls

+ **range:** Southwestern United States

+ **food:** Nuts, seeds, fruit, vegetation, cacti, and invertebrates

Rock Squirrels generally excavate their burrows at a shallower level than those of other ground squirrels, and the burrows also tend to be less extensive. An animal may have a home burrow and another used on feeding forays. The species is social; a dominant male usually is associated with a number of adult females and their young. Mating takes place after hibernation. Females usually have one litter of three to nine young a year in colder areas, and perhaps two litters in warmer habitats with shorter winters. Burrowing Owls and other species that need underground quarters frequently use abandoned Rock Squirrel burrows.

|||||||||||||||||||||||||||||||||||||||||||||||||||||||||||||||||||||||||||||

Spotted Ground Squirrel

Xerospermophilus spilosoma L 9 in (23 cm)

Small and secretive, the Spotted Ground Squirrel is a homebody, spending much time in its burrow when the weather is hot or cold, or with any kind of commotion aboveground.

KEY FACTS

The back is brownish with small pale spots; underside is whitish; the long, narrow tail has a bushy tip.

+ **habitat:** Grasslands, scrublands, and deserts

+ **range:** South Dakota and eastern Wyoming to the southwestern U.S.

+ **food:** Grasses, forbs, cacti, seeds, insects, and small vertebrates

The Spotted Ground Squirrel prefers a habitat with sandy soil that sticks together. Its burrow system can be 15 feet (4.5 m) or more in length with entrances often located under a bush. The species tends to be active in the cooler parts of the day—in the morning and evening— retreating to its burrow during the hottest period. In the colder parts of the range, the species hibernates as early as late July or August. Young of the year tend to linger above into the fall. These squirrels remain underground until April or May. Young are born in the spring in litters of three to five after a gestation of about four weeks.

Thirteen-lined Ground Squirrel
Ictidomys tridecemlineatus L 10 in (25 cm)

The precise patterns of solid lines and dashes on the back of the Thirteen-lined Ground Squirrel make it unmistakable at any age, except at birth. Females produce litters of up to 14 pink babies.

KEY FACTS

Brown body has cream stripes and dashes; cheeks and sides are pale; underside is white.

+ habitat: Pastures, meadows, agricultural land, golf courses, and roadsides

+ range: Central U.S. and south-central Canada

+ food: Seeds, roots, nuts, insects, worms, small vertebrates, and carrion

The Thirteen-lined Ground Squirrel, also known as a gopher, prefers wide-open spaces for its shallow burrow system that may wind some 20 feet (6 m) or more. It doesn't leave mounds around entrances but scatters the dirt far away and plugs side entrances with vegetation. The species undergoes complete hibernation. Its heart rate slows from 200 beats a minute to a mere 4 or 5, and its body temperature can drop to 32°F (0°C). It must awaken before its temperature drops lower or it will freeze to death. Many predators threaten the species, including foxes, Coyotes, snakes, and the badgers that bulldoze through its burrows.

|||||||||||||||||||||||||||||||||||||||||||||||||||||||||||||||||||||||||||

Golden-mantled Ground Squirrel

Callospermophilus lateralis L 11 in (28 cm)

The Golden-mantled Ground Squirrel has a chipmunk-like appearance, with its contrastingly striped sides and bulging cheek pouches that hold hundreds of seeds.

KEY FACTS

Head is orange; back is grayish brown; black-edged white stripe on sides.

+ habitat: Alpine meadows, coniferous and mixed forests, brushy and rocky areas

+ range: Southwestern Canada to California, Arizona, and Utah

+ food: Seeds, nuts, vegetation, fungi, insects, birds, eggs, and carrion

The striped side of the Golden-mantled Ground Squirrel stops short of its head, a feature that distinguishes it from the Western Chipmunk; it also is heavier than the chipmunk. The Golden-mantle spends most of the summer fattening up, and sometimes caching food to prepare for the hibernation it will begin in late summer to mid-fall, depending on climate. The species is also adept at working its good looks for handouts at campsites in the mountain West. This ground squirrel lives solitarily in a burrow that often is excavated at the base of a stump, rock, or log. It curls into a tight ball, head tucked down past its chest.

Alpine Chipmunk
Tamius alpinus L 7 in (18 cm)

The Alpine Chipmunk inhabits rocky areas in a very small range in California's Sierra Nevada, occurring at elevations to almost 13,000 feet (4,000 m).

KEY FACTS

The body is pale over-all, with brown and white stripes on head and side; the sides are orangish, and the underside is whitish.

+ habitat: Alpine meadows and talus slopes from timberline to summit

+ range: Sierra Nevada in California

+ food: Seeds, fungi, berries, eggs, and young birds

Running close to each other at full tilt over rocky mountain slopes, a pair of Alpine Chipmunks displays the species' agility and adaptation to alpine life. This small ground squirrel is about the same size as the Least Chipmunk *(T. minimus),* but its tail is shorter and bushier, and its stripes stop short of the rump. The Alpine may raid the nests of rosy-finches and other birds to feed on eggs and young. But the bulk of its diet seems to be seeds, with which it fills its cheek pouches to near bursting; then it pulls down grass stems and stuffs the seeds into them. It also enjoys what it can beg or steal at campsites.

Cliff Chipmunk

Tamias dorsalis L 9 in (23 cm)

Its sides have stripes, like other chipmunks, but on the Cliff Chipmunk they are dark and often difficult to see. A more distinct black stripe runs down the back.

KEY FACTS

Body is gray with dark mid-back stripe; sides are orangish; face has strong lines; ears are long.

+ habitat: Canyons and other rocky areas with brush, coniferous, and mixed forests

+ range: Southwestern U.S.

+ food: Seeds, nuts, berries, cacti, insects, amphibians, reptiles, birds, and eggs

The Cliff Chipmunk runs agilely across ridgelines and scales canyon walls. It forages from dawn to dusk, filling its cheek pouches with nuts, seeds, and berries to cache or to take to its cliff nest or underground burrow. This desert species faces many predators that attack by land or air, including hawks, Coyotes, badgers, and rattlesnakes. A threat can bring on a period of complete immobilization, which can last from 3 to 25 minutes. A swaying tail signals danger to others. This species has a longer breeding season than most chipmunks, but the female bears only one litter of about four or five young a year.

Merriam's Chipmunk

Tamias merriami L 10 in (25 cm)

A bushier-than-average chipmunk tail is the Merriam's Chipmunk's claim to fame. It may help this California species with climbing, balance, and insulation from the cold.

KEY FACTS

Body is grayish brown to reddish with dark stripes on back; lower, paler stripes are grayish or whitish.

+ habitat: Chaparral, brush, mixed woodlands, and rocky areas to 9,700 feet (3,000 m)

+ range: Central and southern California

+ food: Seeds, nuts, insects, lizards, birds, and eggs

More arboreal than many other chipmunk species, Merriam's Chipmunk frequently dens in stumps, hollow trees, and fallen logs, and raises its young in tree holes. It also may take over the burrows of other species, such as the woodrat. Extremely nimble, it can run up vertical cliffs on its toes and can travel through bushes and trees from branch to branch. It sometimes raids the acorn caches of Acorn Woodpeckers. At mating time, females advertise their readiness by perching in a prominent place and calling repeatedly, using the alarm *chip* sound for 10- to 30-minute periods over the course of three or four hours.

Least Chipmunk
Tamias minimus L 7.5 in (19 cm)

The Least Chipmunk is the most widespread of the western chipmunk species, ranging from alpine tundra to sagebrush flats. It may weigh no more than an ounce or two (28–56 g).

KEY FACTS

Color varies; stripes on head and back; tail as long as body.

+ habitat: Coniferous forests, alpine meadows and tundra, sagebrush, sand dunes, and open areas

+ range: Northwestern and central Canada and much of western U.S.

+ food: Seeds, leaves, buds, flowers, insects, eggs, and carrion

This tiny, high-strung chipmunk is "least" in name only. This highly adaptable species may lodge in an abandoned woodpecker hole, in a nest of leaves and grasses in a tree, or in burrows that it excavates underground. Its coloration varies widely according to region, ranging from very pale with pale stripes in dry areas to reddish brown with distinct stripes in the East. Its feeding habits are equally eclectic, and it stores food that it will eat in its burrow when it periodically wakes up during hibernation. Adults mate shortly after emerging from hibernation; three to six thimble-size young are born in May or June.

Eastern Chipmunk

Tamias striatus L 9.5 in (24 cm)

Eastern Chipmunks travel efficiently, making a beeline across the top of a fence rail or rock wall before disappearing into a well-concealed burrow entrance. Danger elicits a loud, high *chip, chip.*

KEY FACTS

This species is grayish to reddish brown; dark stripe on back; dark-edged white stripe on side; eye bordered in white.

+ habitat: Wood-lands, woodland edges, suburban and urban areas

+ range: Eastern half of U.S.; southeastern Canada

+ food: Nuts, seeds, berries, fungi, insects, worms, and garden bulbs

A rock with a messy pile of crushed acorn shells might signal the outdoor dining terrace of the Eastern Chip-munk. Its burrow has tunnels that may extend more than 100 feet (30 m) and may hold a bushel or more of stored nuts and seeds. Burrows can have a single, large multipur-pose chamber or multiples. Adults live solitarily and are territorial, coming together only for mating. Females may have one or two litters a year, depending upon climate, of three to five young. Overwintering patterns vary by location and from year to year; some chipmunks awaken at intervals to eat food they cached in the fall.

Townsend's Chipmunk

Tamius townsendii L 10 in (25 cm)

One of a number of chipmunks in the Pacific Northwest, Townsend's Chipmunk is a resident of coastal forests, where fallen logs offer den sites and new green plants take hold.

KEY FACTS

Body is tawny, grayish, or reddish brown; stripes are grayish, white, or orange and black.

+ habitat: Moist forests and clear-cuts with fallen logs and new growth

+ range: Oregon and Washington to southeastern British Columbia

+ food: Seeds, fruits, nuts, fungi, lichens, and insects

The Townsend's Chipmunk's day really gets going in the late morning or early afternoon when it is most active. It feeds and gathers food to store in its underground burrow; accumulation of stored food increases in late summer and fall. Some Townsend's Chipmunks will hibernate for the winter, and others in warmer coastal areas will stay active all winter. Still others may exist in an in-between state—present aboveground, but lethargic. Mating occurs in the spring, with litters of three to five born in May or June. A skillful climber, this chipmunk can scurry up a tree to hide or dive into a hole or hollow log to take cover.

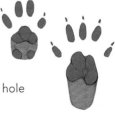

American Beaver

Castor canadensis L 3–4 ft (0.9–1 m)

Nature's master engineer, the American Beaver constantly toils to create its own habitat. By damming streams and building lodges, it transforms the landscape to its needs.

KEY FACTS

Brown coat has coarse guard hairs and dense underfur; tail naked and scaly; hind feet fully webbed.

+ habitat: Streams, rivers, ponds, lakes, and swamps in wooded areas

+ range: Much of the United States and Canada

+ food: Bark, twigs, and leaves of trees and shrubs; herbaceous plants

The physiology of our largest rodent is geared to life in water. Paired orange incisors let it gnaw down trees for building and fell saplings for food. It uses oil from glands at the base of its tail to waterproof its fur. Webbed hind feet aid swimming, and a scaly, paddlelike tail serves as a rudder and as an alarm when slapped on water. Adult beavers form monogamous pairs and live in family groups with three or four kits of the year, as well as yearlings. Beavers along rivers usually do not dam, living in burrows in the banks. Clumsy on land, the beaver is vulnerable to predators such as foxes and Coyotes.

North American Porcupine

Erethizon dorsatum L 26 in (66 cm)

Many a curious dog has returned home whimpering with a face full of porcupine quills. The second largest North American rodent is armed with about 30,000 of them in its head, back, and tail.

KEY FACTS

Black or brown body has yellowish quills partly concealed by long guard hairs.

+ **habitat:** Brushy areas, woodlands, and desert washes

+ **range:** Alaska and Canada into western and northeastern U.S.

+ **food:** Leaves, needles, buds, nuts, fruit, and inner bark of trees; also herbaceous plants

It's an urban legend that the North American Porcupine shoots its quills. It doesn't. Instead, it presents a quill-filled back and tail to a threatening animal; shaking and stomping often causes loose quills to fly off. A lashing tail usually drives quills home. About the only predator able to kill the porcupine without mishap is the Fisher. This weasel relative bites the face until it is able to flip over the porcupine and go for the belly. Porcupines sleep by day in rock crevices, logs, or trees and feed at night on vegetation. Females give birth to a single young with soft, moist quills that quickly dry and stiffen.

Nutria/Coypu

Myocastor coypus L 34 in (86 cm)

This South American native was introduced to the United States for fur farming; many animals escaped or were released deliberately. It is widely considered an undesirable species.

KEY FACTS

Guard hairs of yellowish or reddish brown hide gray underfur; round tail is lightly haired.

+ habitat: Streams, rivers, and freshwater and brackish marshes

+ range: Common in Gulf Coast states; scattered in southeastern and western U.S. and British Columbia

+ food: Aquatic plants and crops

Nearly as big as a beaver but with a narrow, rounded tail, the Nutria has worn out its welcome since its introduction in the 19th century. It decimates wetlands by eating all parts of aquatic plants, laying waste to entire stands of vegetation. It also mows down agricultural fields. The prolific Nutria—a female has up to three litters of up to 12 young a year—is particularly entrenched in Louisiana, which has a population of about 30 million. Newborn Nutrias swim and eat grass within days of birth. They can also nurse while riding on their mother's back, as her mammary glands are on her sides above the waterline.

||||||||||||||||||||||||||||||||||||||||||||||||||||||||||||||||||||||||||

Banner-tailed Kangaroo Rat

Dipodomys spectabilis L 13 in (33 cm)

This southwestern rodent has short front legs and elongated hind limbs, useful for jumping to evade predators. Territorial kangaroo rats fight each other by kickboxing like kangaroos.

KEY FACTS

Back is gray-brown; sides are yellowish brown; underside is white; long, striped tail has a white tip.

+ **habitat:** Short-grass plains, sandy plains, shrublands, and gravelly areas

+ **range:** South-western Texas, New Mexico, and southern and eastern Arizona

+ **food:** Primarily seeds

The Banner-tail is a heavy kangaroo rat, weighing about a third of a pound (150 g). Easily provoked and protective of their seed hoards, two Banner-tails will face off—jumping, twisting, kicking sand, scratching with hind claws, and biting if they get close enough. The species builds elaborate burrows that are enlarged by successive generations. They can be 3 feet (0.9 m) high and 15 feet (4.5 m) wide with multiple entrances, dead-end tunnels, and escape hatches. A storage chamber may hold several pounds of seed, and grass-lined nests shelter three litters of two or three young that may be born from January to August.

Ord's Kangaroo Rat

Dipodomys ordii L 9 in (23 cm)

Like other kangaroo rat species, the Ord's has fur-lined cheek pouches that it stuffs with grass seeds and carries back to its burrow to become part of its hoard.

KEY FACTS

Back varies from grayish to brown to orange; sides are paler; underside is white; crested tail has black tip.

+ habitat: Scrub-lands, piñon-juniper woodlands, brush-lands, and sand dunes

+ range: Southern Alberta and Saskatch-ewan south to Arizona and western Texas

+ food: Seeds and insects

The most widespread kangaroo rat, Ord's is smaller than the Banner-tail and occupies a wider range of habitats. It will drink free water, but it can survive on water that is derived metabolically from its food. Like other kangaroo rats, it must have a nearby source of sand for dust baths to keep its fur sleek and glossy. Ord's is nocturnal, venturing out mainly on moonless nights—but braving the exposure of moonlight when it has to—to collect seeds for its hoard. It burrows under-ground, seldom making an above-ground mound. The species does not hibernate and breeds year-round in some parts of its range.

Plains Pocket Gopher
Geomys bursarius L 11 in (28 cm)

Pocket gophers, also known as "earth mice," are built for burrowing. They have large incisors that are always exposed, small eyes and ears, and large, curved front legs for digging.

KEY FACTS

Back is brown to reddish to black; underside is pale; tail is mostly naked; has large front claws.

+ habitat: Grasslands, roadsides, and croplands

+ range: Central U.S. from northern Texas to Manitoba

+ food: Roots, tubers, and vegetation

The pockets of the Plains Pocket Gopher are the animal's external, fur-lined cheek pouches that hold a haul of roots and tubers for underground storage chambers. Like other pocket gophers, this species lives its life underground. Pocket gophers are also known for their trick of causing an aboveground plant to disappear from view by pulling it into a feeding tunnel. Typically, they travel in shallow feeding tunnels and harvest the food dangling into them. Living quarters are deeper and contain a nest chamber, toilet, and food storage. Pocket gophers are solitary, except at mating time, and resent intrusion into their turf.

California Pocket Mouse

Chaetodipus californicus L 8 in (20 cm)

Pocket mice, like pocket gophers, have external cheek pouches used to transport seeds to their burrows. Some two dozen species live in the western United States; several extend into Canada.

KEY FACTS

Back is grizzled brown, sides are paler; distinct orange lines separate the sides and whitish underside; ears are long.

+ **habitat:** Chaparral, grasslands, scrublands, and woodlands

+ **range:** Central and southern California

+ **food:** Seeds, insects, and vegetation

The California Pocket Mouse lives mostly underground and alone. It remains in its burrow during the day with entrances plugged to keep up humidity below. It comes out at night to forage, then returns to the burrow to deposit its take, using paws to push out the seeds in its cheek pockets, sometimes turning the pockets inside out. The species is picky about the weather: It will become torpid and stay underground if it is too hot or too cold out, or if the weather is inclement. It does not enter true hibernation. Wildfires can kill most of an area's population, but the burned area usually is only temporarily vacated.

Woodland Jumping Mouse

Napaeozapus insignis L 8.5 in (22 cm)

The woodland counterpart to the Meadow Jumping Mouse, this species has the characteristic long tail, differentiated by a white tip. The sides of the woodland species show more color contrast.

KEY FACTS

Back is dark brown; sides bright orange or reddish brown; underside is white; white-tipped long tail.

+ habitat: Wet coniferous and mixed woodlands, stream banks, and swamp edges

+ range: Southeastern Canada and northeastern U.S. to Appalachians

+ food: Seeds, berries, fungi, and insects

The Woodland Jumping Mouse lives up to its name, preferring woodland and sheltered habitats to open ground. Meadow Jumping Mice are much more likely to enter the woodland mouse's habitats than the reverse, but the two species are not considered to be competitive with each other. Woodland Jumping Mice dig their own burrows or appropriate abandoned ones, preparing a nest for a long hibernation of about eight months. Like the meadow mice, they accumulate fat for hibernation, but do not store food. Three to six young, born in late spring or summer, practice jumping that usually is measured in inches as early as three weeks.

Meadow Jumping Mouse

Zapus hudsonicus L 9 in (23 cm)

The Meadow Jumping Mouse is a "bigfoot" species. Oversize rear feet on long legs give it the propulsion it needs to make jumps as high as 6 feet (1.8 m) to elude predators.

KEY FACTS

Body is dark brown with yellowish brown sides; underside is whitish; hind feet are one-third of body length; extremely long tail.

+ habitat: Fields, meadows, stream banks, and marshes

+ range: Southern Alaska and Canada into central and eastern U.S.

+ food: Fungi, fruit, and insects

In addition to big feet, jumping mice have long tails and tricolor coats tailored to a species' particular habitat. The nocturnal Meadow Jumping Mouse's colors of dark brown, yellow, and white blend with a meadow's colors at dusk and dawn. When a predator homes in on a mouse, it hops erratically in an evasive zigzag fashion. In summer this mouse nests on the ground or in shallow burrows. Two litters of three to seven young may be born in one year. In winter it builds a deeper burrow for hibernation; often alone or sometimes paired, it tucks up into a tight ball that cannot quite accommodate those oversize feet.

Northern Collared Lemming

Dicrostonyx groenlandicus L 5 in (13 cm)

Northern Collared Lemmings are one of the few mammals that molt to a snowy white winter coat. In Inuit legend, the white animals cascaded from the sky with snowflakes.

KEY FACTS

Back is gray in summer, orangish on throat and sides; underside is white; dark stripe extends from nose.

+ **habitat:** Rocky areas and meadows on Arctic tundra

+ **range:** Arctic from Alaska through Canada, and western edge of Greenland

+ **food:** Tundra vegetation, primarily forbs and willow

This tiny rodent is active year-round, day and night, in the challenging Arctic environment. It builds short burrows in summer; females modify these burrows to prepare for nesting. As winter nears, the forefeet of this lemming transform into tools for burrowing into snow and ice: Two middle claws lengthen, and claw pads enlarge and toughen. In spring, the feet return to their former shape.

Winter nests are lined with grasses and sedges and are insulated by the snow. Lemming numbers fluctuate in four-year cycles related to the weather and the food supply. At population peaks, some undertake short migrations in search of food.

Ungava Collared Lemming

Dicrostonyx hudsonius L 5 in (13 cm)

This lemming inhabits a peninsula in northern Quebec and Labrador, which lends it the name Ungava. It undergoes the same seasonal fur changes as the Northern Collared Lemming.

KEY FACTS

Back is dark grayish brown with black stripe along spine; throat and sides orange; underside is gray.

+ **habitat:** Tundra, rocky hillsides, and meadows

+ **range:** Ungava Peninsula of northern Quebec and Labrador and islands to the west

+ **food:** Willow, birch, and aspen twigs

In their winter fur, Ungava Collared Lemmings, like the other collared lemmings that molt to white for the season, tend to look like short, stocky, white puffballs. Their forefeet also transform to digging claws in preparation for winter excavations. The range of this species abuts, but does not overlap with, that of the Northern Bog Lemming. The Ungava has the same system of short summer burrows and winter tunnels, and nests under the snow. The small range of this species is a remnant of its former presence, going back tens of thousands of years to the Pleistocene, when it inhabited the North American and Asian Arctic.

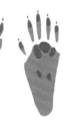

Brown Lemming
Lemmus trimucronatus L 5 in (13 cm)

Many animals of the tundra—such as foxes, weasels, and owls—rely on the Brown Lemming for sustenance. In years when the rodent is scarce, predator populations suffer.

KEY FACTS

Back and sides are yellowish brown; rump is cinnamon; has orange tufts at base of ears; tail is very short and concealed.

+ habitat: Low wet tundra with ample vegetation

+ range: Alaska through far northern Canada and into British Columbia

+ food: Grasses, sedges, and mosses

Predators feed on Brown Lemmings, and the lemmings feed on grasses and sedges in summer and mosses in winter. When they have depleted an area's vegetation, lemming breeding crashes. In peak lemming years, some of the rodents emigrate into towns and across bays and lakes to survive. Lemmings are typically precocious and prolific reproducers. Females can begin breeding as young as two weeks old. They breed all winter, even under the snow, and can have as many as three litters a year. In one documented case, a pair produced eight litters in 167 days—an average of a litter every 20 days—and then the male died.

Rock Vole/Yellow-nosed Vole

Microtus chrotorrhinus L 6 in (15 cm)

The elusive Rock Vole lives in scattered populations within its range. Its profile shows a more mouse-like appearance than that of other voles.

KEY FACTS

Back is brown; under-side is gray; nose area is yellowish orange.

+ **habitat:** Talus slopes and other rocky areas with hardwood or mixed forests near water

+ **range:** Eastern Canada and northeastern U.S. to Appalachians

+ **food:** Bunchberries, blackberries, other vegetation, fungi, and insects

The Rock Vole is well named for its habitat, as it is for its colorful nose as an alternate common name, the Yellow-nosed Vole, and species name, *chrotorrhinus* (colored nose). The species is active year-round, day and night, spending a good part of the time in underground runs. It builds nests of plant fibers and mosses that are often tucked into rock crevices. A separate chamber is used for a latrine. The species breeds during much of the year; a female may have multiple litters of two to five young. In parts of its range, the Rock Vole shares its habitat with the Red-backed Vole, but the two do not seem competitive.

Prairie Vole
Microtus ochrogaster L 6 in (15 cm)

Runways are central to a vole's life, especially to a Prairie Vole. The rodents clip grass short and tamp down a path from repeated use. Longer grass on the sides helps to hide the passageways.

KEY FACTS

Body usually has long, grizzled, grayish brown fur above; sides are paler; underside is cream or whitish.

+ **habitat:** Prairies, grasslands, and agricultural areas

+ **range:** Central U.S. into south-central Canada

+ **food:** Seeds, roots, tubers, flowers, leaves, insects, and crops

In addition to extensive aboveground runways, Prairie Voles dig connected underground tunnels that include chambers for food caches and lined nests, which may be concealed under logs. The voles are active year-round and rely on stored roots and tubers in winter. Prairie Voles are unusual among rodents for their monogamy; once mated, they seem to mate for life, short as it is. The males also help with the care of their offspring, which can be born at any time of year, but most often in summer in litters of three to five. The mates stand together against the encroachment of other voles into their territory.

Meadow Vole
Microtus pennsylvanicus L 6.5 in (16.5 cm)

The Meadow Vole is a linchpin in the vertebrate food chain, supporting species as varied as birds, bears—even fish, which snap up swimming voles. All this makes for a short average life span.

KEY FACTS

This vole has thick fur on its dark brown back; its sides are paler; the underside is whitish; the tail is long.

+ habitat: Meadows, roadsides, grasslands, and orchards

+ range: Alaska through Canada and into northern and central U.S.

+ food: Green plants, roots, tubers, and bark

Meadow Voles may be the go-to lunch or dinner for many species, but when conditions are favorable, their reproduction rates can explode. The species may breed year-round, and a female usually produces five or six young in each litter; in years of population boom, an acre of meadow can house hundreds if not thousands of voles. They can strip a farm field, grazing like hoofed mammals with teeth adapted for grinding. They often sit on their haunches and hold food daintily in their forepaws. Meadow Voles construct grass-roofed runways as transit paths, and they nest in the grass or in burrows they dig with many entrances.

Woodland Vole/Pine Vole

Microtus pinetorum L 5 in (13 cm)

The burrowing evidence of the beady-eyed, short-legged, and short-tailed Woodland Vole looks like the work of a mole. This velvet-furred vole sometimes forages aboveground.

KEY FACTS

Body is auburn with grayish underside; ears and eyes are tiny; tail is very short.

+ habitat: Hardwood forests with thick substrate, clearings, and orchards

+ range: Central and eastern U.S. and slightly into southeastern Canada

+ food: Roots, tubers, grasses, seeds, fruits, and bark

The Woodland Vole's smooth, molelike coat helps it move easily underground and shed soil. Active day and night, the vole digs shallow tunnels, leaving a wake of ridges along the ground. It can excavate up to 15 inches (38 cm) in a minute. This vole's feeding habits frustrate gardeners and growers: It can devastate a potato crop and kill apple trees by eating roots and gnawing bark. These unpopular activities have earned it the nicknames "potato mouse" and "apple mouse." The species was first scientifically described from a specimen in the Georgia pines and named *pinetorum;* however, it is found more often in hardwood settings.

|||||||||||||||||||||||||||||||||||||||||||||||||||||||||||||||||||||||||||||||||||||||||

Southern Red-backed Vole

Clethrionomys gapperi L 5.5 in (14 cm)

The Southern Red-backed Vole is a resident of cool, damp forests, white cedar swamps, mountain ridges, and ferny glades. A broad chestnut stripe runs from its forehead to the base of its tail.

KEY FACTS

A chestnut stripe extends along the head and back; sides are grayish; underside is silvery or cream.

+ habitat: Forests, clearings, mountain meadows, and swamps

+ range: Canada and the northern U.S. into Rockies and Appalachians

+ food: Seeds, nuts, berries, fungi, lichen, insects, and carrion

Southern Red-backs travel on runways built by other rodents. The high-strung vole constructs small, spherical nests woven of grass, leaves, and moss—or at times, simple platforms—under logs or tucked into tree roots. In many areas, this species may be grayish with a sooty dorsal stripe. It is active year-round and can be nocturnal or diurnal, depending on the season. Red-backs breed about eight months of the year; females have two or three litters of four to eight young. The Northern Red-backed Vole (*C. rutilis*), a similar species that occurs in tundra and taiga habitats, has a more orange back and a hairier tail.

Round-tailed Muskrat/Florida Water Rat
Neofiber alleni L 13 in (33 cm)

The Round-tailed Muskrat makes smaller and more delicate conical homes than the Muskrat. It also has a round tail, not a laterally flattened one like the larger Muskrat.

KEY FACTS

The body is glossy brown above with buff underside, and the tail is almost hairless.

+ habitat: Freshwater marshes and agricultural areas

+ range: Throughout much of Florida and southeastern Georgia

+ food: Aquatic plants, including grasses and water lilies

The Round-tailed Muskrat seeks shallow, watery habitats with abundant aquatic plant life and fashions its round home on the surface with two underwater entrances. These muskrats densely populate favored areas; in central Florida, an acre can hold up to 120 adult Round-tails. The animals live individually, but may share feeding platforms—elevated pads of vegetation. The species breeds year-round; the female gives birth to one to four young in the "house," which she prepares with a lining of fine, dry grasses. Round-tails succumb to a variety of different predators, including hawks, owls, herons, snakes, and Bobcats.

Muskrat

Ondatra zibethicus L 21 in (53 cm)

This mammal exudes musk, but it is not a rat. Rather, the Muskrat is more like a large amphibious vole, or field mouse, which seldom strays from water.

KEY FACTS

Body is glossy brown above; grayish on underside; ears are tiny; hairless tail is flattened.

+ habitat: Freshwater and brackish marshes, streams, and ponds

+ range: U.S. and Canada, except arid Southwest and extreme Southeast

+ food: Aquatic plants, crayfish, fish, turtles, and mollusks

Muskrats share habitat with North American Beavers, and the two are often confused. In addition to being smaller than a beaver, the Muskrat has a naked tail and when it swims, more of its body appears above water than a beaver's. Sometimes it shows two humps in addition to the head, for a kind of Loch Ness Monster silhouette. Muskrats build their lodges out of cat-tails and marsh grasses, not trees. They also build tunnels, complex runways, and stream-bank burrows. They mate multiple times in a year; a female may bear up to eight litters of four to eight young. Trappers take millions of this prized fur species every year.

Western Heather Vole
Phenacomys intermedius L 5.5 in (14 cm)

This is the heather vole of higher western elevations, which does not have orange coloration on its nose and ear tufts. It shares many habits with the Eastern Heather Vole.

KEY FACTS

Back and sides are grayish to reddish brown; underside is white; tail is short.

+ habitat: Upland forests, dwarf shrubs, alpine meadows, and rocky areas

+ range: Southwestern Canada to California and Rockies to New Mexico

+ food: Shrub bark, berries, seeds, leaves, and lichens

The Western Heather Vole is usually smaller than the eastern species, and its very long whiskers extend well back beyond its ears. In the manner of the eastern species, it nests underground in the summer (caching food by the entrance) and aboveground under the snow in winter, another shared trait. Winter nests are made of grasses and may be sealed with lichens and set at the base of a shrub, rock, or stump. A family group of females and young may nest for warmth in the winter. The species otherwise is solitary and aggressively territorial. A female may bear several litters of two to eight young in spring and summer.

Eastern Heather Vole

Phenacomys ungava L 5.5 in (14 cm)

Until recently, the Eastern and Western Heather Vole were considered a single species, *P. intermedius.* Differences include the orange hairs on the Eastern's nose and in its ears.

KEY FACTS

Back and sides are brown; top of nose and tufts in front of ears are orange; tail is short.

+ habitat: Heather and other shrubby areas, coniferous forests, wet meadows, and woodland edges

+ range: Most of Canada and into northeastern Minnesota

+ food: Shrub bark, berries, and vegetation

Although it occupies a vast continental range, the Eastern Heather Vole, also called the Ungava Vole, seems to live in somewhat isolated populations. It is known to be very elusive, and it is noted for caching food that it gathers at the entrance to its burrow at night for next-day consumption. The species does not hibernate. It digs short burrows, mostly for use in summer, that contain nests of grasses, leaves, and lichens. In the winter it nests under snow. Females may have as many as three litters of between two to eight young each year. The species often falls prey to hawks, owls, and weasels.

Southern Bog Lemming

Synaptomys cooperi L 5 in (13 cm)

The Southern Bog Lemming is a resident of spongy bogs and woodland glades, usually near a source of grasses and sedges. It may live temporarily in scattered small colonies.

KEY FACTS

Back is grizzled brown, sides are yellowish brown; underside is silvery; tail is short.

+ habitat: Bogs, meadows, coniferous forests, and clear-cuts

+ range: Southeastern Canada and central and eastern U.S.

+ food: Grasses, sedges, berries, fungi, moss, and bark

Like many voles and other lemmings, Southern Bog Lemmings travel on grassy runways. They also burrow into the ground, adding chambers for feeding, caching food, and resting to the tunnels they dig. Grass clippings about an inch long may be stacked next to burrow entrances. Unlike other lemmings and their rodent cousins, bog lemmings produce bright green scat, a distinctive sign of their presence. The similar Northern Bog Lemming (*S. borealis*) tends to be larger and has orange fur at the base of its ears. It ranges from Alaska through northern Canada to the northeastern U.S. and across the borders of some western states.

Bushy-tailed Woodrat

Neotoma cinerea L 16 in (41 cm)

The Bushy-tailed Woodrat is a western species that makes itself a nuisance to cabin dwellers from the Yukon to Arizona, shredding furnishings and leaving a musky urine trail.

KEY FACTS

Back color varies from grizzled yellowish gray to black; sides are orange or yellowish; underside is whitish; tail is bushy.

+ habitat: Rocky areas, caves, canyons, and cliffs

+ range: Western Canada to California, Arizona, and the Dakotas

+ food: Vegetation, fruit, nuts, seeds, and fungi

This highly territorial large woodrat makes stick-and-vegetation nests festooned with collected materials, droppings, and food leavings. It anoints the nest with musky urine to establish ownership. Over time the urine hardens the structure, creating a midden that can last thousands of years, providing valuable scientific information. Inside the nest is a larder of seeds, nuts, and other foods to sustain the nonhibernating rodent through the winter. The Bushy-tail nests in abandoned mines, cabins, caves, and cliffs, from sea level to 14,000 feet (4,300 m). Three or four young are born each spring, and a second litter often follows.

Eastern Woodrat
Neotoma floridana L 15 in (38 cm)

Woodrats are "pack rats," those rodents whose habits are used to describe humans who hate to let any possession go. Pack rats construct houses of natural materials and collected "stuff."

KEY FACTS

Body above is grayish, dark brown, or sandy; underside and tops of feet are white; tail is long and haired.

+ habitat: Woodlands, slopes, plains, scrublands, and rocky areas

+ range: Mainly eastern and south-central U.S.

+ food: Seeds, nuts, grasses, buds, berries, and fungi

The wide-ranging Eastern Woodrat forages mainly at night, aided by its large eyes and long whiskers. It constructs a bulky house, perhaps 3 feet (0.9 m) tall and 5 feet (1.5 m) wide, out of sticks mounded haphazardly in a rock crevice, on the ground, or in a tree. At the nest, it incorporates "found" materials, ranging from jewelry to cooking utensils to small car parts. Sometimes when carrying an object it will find another that it prefers and drop its first find for the other. This has led to the moniker "trade rat," and sets up tall tales; in one tale, a woodrat stole a 50-cent piece and left two quarters behind.

Golden Mouse
Ochrotomys nuttalli L 6 in (15 cm)

If mice had beauty contests, the Golden Mouse would be a contender with its soft, burnished orangey coat. It has talent, too: It can hang like a monkey from its prehensile tail.

KEY FACTS

Body above is a golden orange-brown; sides are paler; underside is yellowish.

+ **habitat:** Woodlands with brush, vines, and briers

+ **range:** Southeastern U.S., except extreme southern Louisiana and southern Florida

+ **food:** Seeds, nuts, berries, and invertebrates

The Golden Mouse once was grouped in the genus *Peromyscus*, with species such as deermice, before it was placed in a genus of which it is the sole member: *Ochrotomys*, meaning "gold mouse." This species uses its climbing skills to evade predators and flooding. The primarily nocturnal and arboreal mouse carries seeds in its cheek pouches up vines and branches to feeding platforms. Multiple platforms are in a nest, a globular structure about 7 inches (18 cm) in diameter. Several adults may occupy a nest until breeding time, when a female with a litter of two or three young displaces all other adults, including her mate.

Northern Grasshopper Mouse

Onychomys leucogaster L 6 in (15 cm)

True to its name, the short-tailed Northern Grasshopper Mouse relishes a grasshopper meal. It is typically bigger than the southern species.

KEY FACTS

Body color ranges from gray to brown to orangish; underside is white; individuals may have white tufts at base of their ears.

+ habitat: Deserts, grasslands, and shrublands

+ range: Western U.S. to Saskatchewan and Alberta

+ food: Insects and other arthropods, lizards, mice, seeds, and vegetation

Unusually for rodents, grasshopper mice are bona fide predators, stalking and killing not only insects such as grasshoppers, but other live prey such as scorpions, lizards, and even other mice. Their bodies have adapted to a carnivorous diet with its built-in infusion of water, but these species also eat seeds and vegetation when hunting is poor. The Northern Grasshopper Mouse makes nests in burrows that it digs or appropriates from other animals. The father participates in rearing two to six young in several litters a year. The species fiercely defends a territory of about 7 acres (2.8 ha) from other grasshopper mice.

Southern Grasshopper Mouse

Onychomys torridus L 5.5 in (14 cm)

Similar in many ways to the Northern Grasshopper Mouse but smaller, the southern species makes its home in southwestern arid shrublands at low elevations.

KEY FACTS

Body is gray to cinnamon to pinkish; underside is white; tail is longer than half its body length.

+ habitat: Desert shrublands at low elevations

+ range: Southwestern New Mexico and southern Arizona to southern Nevada and California

+ food: Insects, other arthropods, mice, and seeds

The grasshopper mouse frequently ambushes other mice from behind, dispatching them with a bite to the head, weasel-style. As it eats, this mouse may rear up on its hind feet, lift its head, and let out a high-pitched howl. The howl may be a territorial warning, which is typical of these vocal rodents. The Southern Grasshopper Mouse does not have all the physiological adaptations of other desert animals, but it is able to obtain water from its diet. Also known as the Scorpion Mouse, it has developed a method of avoiding a venomous sting while killing a scorpion by immobilizing the head before it bites the tail.

Brush Deermouse

Peromyscus boylii L 7.5 in (19 cm)

The Brush Deermouse, like the Piñon Deermouse, belongs to the genus *Peromyscus*, noted for difficulty in identification. Within it, the western species are especially challenging.

KEY FACTS

Back is grayish brown; sides are orange; underside is white; ears are large.

+ habitat: Piñon-juniper woodlands and chaparral with rocky areas, brush, and logs

+ range: Often at elevations above 2,000 feet (600 m) in the western U.S.

+ food: Seeds, nuts, fruit, leaves, fungi, and cacti

The Brush Deermouse gravitates to habitats with a lot of cover in the form of brush and rocks it can use for shelter and nesting. The species also infiltrates cabins in the West, where it is the most abundant *Peromyscus* in mountain regions. Genders come together only at mating time, and evidence suggests that multiple males can father the offspring in one litter, which usually numbers between two and five young. The main difference between the Brush Deermouse and the Piñon Deermouse is the latter's larger ears. Comparing ears with the size of the hind feet aids identification: The ears of these mice are usually not longer than the feet.

White-footed Deermouse

Peromyscus leucopus L 6.5 in (16.5 cm)

The White-footed Deermouse and its almost twin, the North American Deermouse, tend to confuse people, but subtle differences exist. The two species do not interbreed.

KEY FACTS

Back is dark brown; sides are orangish; underside and feet are white.

+ habitat: Woodlands, brushlands, fields, cleared areas, and riparian areas

+ range: Central and eastern U.S., except in extreme Southeast

+ food: Nuts, seeds, insects, vegetation, and fruit

A whiter underside and a slightly furred tail are two of the physical differences that distinguish the White-footed Deermouse from the North American Deermouse. Although the two species share a lot of range, the White-foot often prefers woodlands, but it also thrives in the Southwest. It swims well, a skill that extends its distribution to islands. Breeding may take place year-round, and the average litter has four young, which may be born in a tree nest, a stone wall, or an appropriated burrow. Food is often cached nearby. These abundant mammals barely live a year and are prey to many aerial and terrestrial predators.

North American Deermouse

Peromyscus maniculatus L 7 in (18 cm)

Superabundant and widespread North American Deermice make their home in nearly every kind of North American habitat, ranging from sea level to 11,500 feet (3,500 m).

KEY FACTS

Back is pale gray to reddish brown; underside is white; tail is furred.

+ habitat: Forests, grasslands, deserts, tundra, swamps, and mountains

+ range: Most of U.S. and Canada, except southeastern U.S.

+ food: Seeds, nuts, berries, fungi, insects, and other invertebrates

This nocturnal deermouse forages everywhere for the many items in its diet. Deermice do not hibernate, but groups may huddle together for warmth in cold weather. Unlike many other rodent species, females may permit their mates to stay in the nest and help with the litter of about four young. Sometimes she will leave him to care for dependent young while she moves to a new nest for her next delivery. Nests can turn up anywhere—in logs, stumps, abandoned burrows, sheds—or even an unused coat pocket. This species, along with the White-footed Deermouse, may host the Deer Tick, which can transmit Lyme disease.

Piñon Deermouse

Peromyscus truei L 7.5 in (19 cm)

The Piñon Deermouse has big feet and even bigger, almost comically large ears. It prefers a rocky habitat and an abundant supply of pine nuts and juniper seeds.

KEY FACTS

Back is gray or brown, sides are orangish; underside is white.

+ habitat: Arid, rocky areas with piñon, juniper, pine, and chaparral

+ range: Western U.S. from west Texas to California, Oregon, and southern Idaho

+ food: Seeds, nuts, leaves, insects, spiders, and fungi

The Piñon Deermouse sleeps by day and forages by night. It nests in hollow trees and rock crevices and often stores its surplus of seeds. Rocky slopes and cliffs are not deterrent to this species—also known as the Big-eared Cliff Mouse—and it ranges to about 10,000 feet (3,000 m). This species can be distinguished from the similar Brush Deermouse by its longer ears and usually larger size. Mortality for these mice is very high. Many factors, including predation from hawks, foxes, and Coyotes, conspire to allow only one in a litter of five or six young to reach a year of age. Females may have several litters a year.

Eastern Harvest Mouse

Reithrodontomys humulis L 4.5 in (11 cm)

The genus name for this very tiny mouse—*Reithrodontomys*—means "groove-toothed mouse"; one of its distinguishing features is a groove that runs down the front of each incisor.

KEY FACTS

Back is reddish brown; sides are paler; underside is grayish; long tail is nearly hairless.

+ **habitat:** Meadows, old fields, thickets, ditches, and marshes

+ **range:** Southeastern U.S. north to Ohio and west to eastern Texas

+ **food:** Seeds of grasses and forbs and insects

The Eastern Harvest Mouse brings home its gleanings of seeds to store. It lives snugly with a mate in a spherical nest as small as 3 inches (7 cm) in diameter, which it weaves from plant materials and attaches to a plant stem about a foot (30 cm) or more off the ground. The nest is lined, often with thistledown or cattail fuzz. Two to seven young are born in the nest from spring to fall, or year-round in the southern range. The similar Plains Harvest Mouse *(R. montanus)* is a bit larger and has a narrow, dark stripe down its back. Its south-central U.S. range slightly overlaps that of the Eastern.

Marsh Rice Rat
Oryzomys palustris L 9 in (23 cm)

The Marsh Rice Rat flocks to fields of newly planted rice to avail itself of the thoughtfully provided seeds. If the field is flooded, no problem—this rodent is an accomplished swimmer.

KEY FACTS

Body is brown to black with lighter sides; underside is whitish; tail is bicolored.

+ habitat: Marshes, riverbanks, wet meadows, and other wet areas

+ range: Atlantic coast south from New Jersey; Gulf Coast to Texas; inland to Kansas

+ food: Invertebrates; fish, turtles, vegetation, and carrion

The Marsh Rice Rat, one of two rice rat species that occur in the United States, makes its home in coastal and inland marshes, riverbanks, canebrakes, and wet meadows, foraging both on land and in water. In addition to seeds, these rodents enjoy a diet of shellfish, other invertebrates, fish, and turtles. This species weaves grassy nests placed on high ground, under a log, or perhaps in a tangle of weeds. In flood-prone lands the nests may hang above high water on a cattail stem or in a muskrat lodge. Marsh Rice Rats also build feeding platforms from bent grasses and connect them by trails to their nests and water sources.

Hispid Cotton Rat

Sigmodon hispidus L 10 in (25 cm)

The Hispid is the most widespread of the cotton rat species,
a group of stocky vole-like rodents. "Hispid" refers to the animal's
rough coat and "cotton rat" to its attraction to cotton crops.

KEY FACTS

Back is grizzled buff
or gray; eastern popu-
lations are darker than
western.

+ habitat: Grass-
lands, fields, agri-
cultural lands, and
scrublands

+ range: Southeast-
ern and south-central
U.S. to Arizona and
southern California

+ food: Grasses and
other vegetation,
insects, birds, and
eggs

Agricultural development in the South provided ideal
circumstances for the Hispid Cotton Rat, a mainly
tropical species, to extend its range and expand its population to the point that it is often the most abundant mammal wherever it is found. These cotton rats create networks
of long and broad runways in the grass, which lead to
nests in sheltered spots on the surface
or in shallow burrows. Females may
breed year-round, starting as early
as six weeks old, and are very prolific.
Litters average five or six young, but
may be as high as 14. On the other hand,
mortality is also high as a result of drought
and predation.

House Mouse

Mus musculus L 6.5 in (16.5 cm)

Star of stage, screen, television, and literature, the House Mouse is a species that has been imported worldwide from Asia. It lives everywhere and eats nearly everything.

KEY FACTS

Body is brownish gray above; underside is buff; tail is long and thick.

+ habitat: Diverse, including meadows, grain fields, sand dunes, and buildings

+ range: Entire U.S. and southern Canada

+ food: Seeds, grain, insects, human food, and some nonfood household products

The House Mouse adapts to—and thrives in—most environments. Its small size and ability to squeeze through very narrow crevices gives it an "all-access pass" to anywhere it wants to go. House Mice live in colonies headed by a dominant male, several females, and their offspring. The females are potential breeding machines; they can produce up to 14 litters a year of 3 to 12 young, enduring a perpetual state of pregnancy. Mouse infestation in a building usually is announced by a musky odor, probably from urine used for territorial marking. This species has been bred into a white mouse form used in scientific experiments.

Norway Rat/Brown Rat

Rattus norvegicus L 16 in (41 cm)

In biology and behavior, the Norway Rat is consummately adaptable, leading to its unmitigated success everywhere it goes—which is everywhere. A superb swimmer, it owns the sewers.

KEY FACTS

Back is yellowish brown; underside is grayish or yellowish; tail is not as long as head and body.

+ habitat: Mainly suburban and urban areas; also fields and marshes

+ range: Throughout U.S. and populated areas of Canada

+ food: Grains, seeds, fruit, chickens, and garbage

The Norway Rat can live comfortably in the country, but it prefers a human connection, swarming to areas of human density with their endless supplies of food waste and diversity of shelter options. This species is a burrower, living on ground floors and in basements, tunnels, compost, and garbage dumps. The Norway Rat eats a wide range of foods but approaches new foods with caution and is wary of poisoned baits. It lives hierarchically, with dominant males and their females and offspring afforded closest access to the food supply. Successful breeding has created many individuals with strong resistance to anticoagulant poisons.

Black Rat/Roof Rat

Rattus rattus L 14.5 in (37 cm)

Unlike the burrowing Norway Rat, the Black Rat lives in lofty places, preferring attics and rafters of buildings and even living in trees in favorable climates.

KEY FACTS

Body is dark brown or black above; underside is grayish or whitish; tail is longer than head and body.

+ habitat: Buildings, ships, dockside structures, coastal forests, and fields

+ range: Coastal and southern U.S., West Coast to British Columbia

+ food: Grains, seeds, fruit, and garbage

The Black Rat colonized North America so effectively in large part by disembarking daily through ports in rat-infested ships, taking up residence in wharves and warehouses, and then dispersing to both urban and rural areas. Modern shipbuilding techniques have basically blocked that immigration route. This rat can lay waste to agricultural crops and has a detrimental effect on native species. The Black Rat is smaller and less stocky than the Norway Rat, and as an adult it lives solitarily, except for females and their newborns and at times even weaned offspring. This rat can breed year-round, producing litters of four to ten.

American Pika/Cony

Ochotona princeps L 8 in (20 cm)

The American Pika, also called a Cony, lives above the timberline in the West. Shaped more like a guinea pig than a rabbit, it is the only lagomorph (includes rabbits and hares) that caches food.

KEY FACTS

Body is brown, orangish, or grayish with orangish sides of neck; ears are rounded; tail is very small; feet are fully furred.

+ **habitat:** Rocky areas, often above 8,000 feet (2,400 m)

+ **range:** British Columbia and Alberta to U.S. western mountain states

+ **food:** Grasses, sedges, and forbs

Many aspects, down to its mouselike scurrying and diurnal lifestyle, distinguish the American Pika from rabbits and hares. During the summer, the Pika harvests vast amounts of grasses that it spreads out to dry before stacking them in large piles for the winter. This normally fairly tolerant animal fiercely guards its hoards against potential poachers. To make the hoards less desirable, it marks them with urine and scent from a facial gland. Like other lagomorphs, the Pika reingests its fecal pellets. It communicates by barking and whistling. The similar Collared Pika (*O. collaris*) has a more pronounced neck ring.

Pygmy Rabbit
Brachylagus idahoensis L 10.5 in (27 cm)

This tiny species was once placed in the same genus with cottontails, but now occupies its own, *Brachylagus*, which means "short hare." Its tail is gray all over, lacking the cottontail look.

KEY FACTS

Body is dark gray in winter, paler in summer; underside is whitish; ears are heavily furred; small tail is gray.

+ habitat: Areas with dense sagebrush

+ range: Great Basin and intermountain areas of western states, with an isolated endangered population in Washington

+ food: Mainly grasses and sagebrush

The Pygmy Rabbit depends heavily on sagebrush for both protective cover and for a large portion of its nutrition. It comes out mainly at dawn and dusk to feed on scrubby, arid-land shrub and uses it to help hide its warren. Unlike most North American rabbits and hares, the Pygmy Rabbit does not nest in a den fashioned in vegetation, known as a form, but in a warren that it digs or appropriates—and then renovates—from another burrowing animal. It tends to stay close to home, and if it is alarmed while away from the nest, it scurries to one of its many warren entrances that are hidden under bushes from predators.

Antelope Jackrabbit

Lepus alleni L 25 in (64 cm)

The enormous ears of the Antelope Jackrabbit can reach up to 8 inches (20 cm) long—a quarter of the animal's body surface, helping it to cool off in the intense heat of its southwestern range.

KEY FACTS

Back is grizzled buff; sides are gray; underside is white; ears have white edges and tips; tail is black above and white below.

+ habitat: Dry grassy areas, with shrubs and cacti, to 5,000 feet (1,500 m)

+ range: Southern Arizona

+ food: Grasses, leaves, including mesquite, and cacti

Only one native mammal can outrun the Antelope Jackrabbit: the Pronghorn, a hoofed species sometimes mistakenly called an antelope. The namesake hare has been clocked at more than 40 miles (64 km) an hour, covering more than 20 feet (6 m) in a single bound. It also adopts a gait in which the hind feet hit the ground several feet ahead of where the forefeet left the ground, varying its path in a way that frustrates predators. In escaping, it displays a flashing patch of white hair on its rump, which is controlled by its back muscles. Active at dawn and dusk, and at night, the species rests camouflaged during the day.

Snowshoe Hare

Lepus americanus L 17 in (43 cm)

This northern hare changes coats with the seasons, from browns and buffs in summer to white in winter, a trait that gives it the alternate name of Varying Hare.

KEY FACTS

Body is brown or red-dish brown, white in winter; short ears are tipped in black.

+ habitat: Brushy areas, coniferous and hardwood forests

+ range: Alaska through Canada to northern U.S., western mountains, and Appalachians

+ food: Vegetation, berries, twigs, and bark

Not much larger than a cottontail and the smallest member of its genus, the Snowshoe Hare is distinguished by large, broad hind feet matted on the bottom with coarse hairs. The feet make it nimble in deep snow, able to bound away quickly under threat. Its white winter coat provides camouflage in a snowdrift; only dark eyes and black-tipped ears betray its presence. Females are a bit larger than the males and may have up to five litters a year of one to nine young, known in their first year as leverets. Snowshoe Hare populations fluctuate in ten-year cycles, perhaps caused in part by disease transmission due to overcrowding.

Arctic Hare
Lepus arcticus L 24 in (61 cm)

North America's heftiest hare, weighing up to 12 pounds (5.4 kg), the Arctic Hare makes its home on the frozen tundra of the Canadian high Arctic.

KEY FACTS

Body is grayish brown in summer, white in winter; always black-tipped ears are shorter than its head.

+ habitat: Treeless tundra, often in rocky areas

+ range: Canadian Arctic mainland and islands; coastal Greenland

+ food: Dwarf willow and other vegetation, berries, moss, and lichens

The Arctic Hare comes well equipped for life in a challenging habitat. Its chunky body bears thick, woolly white fur that molts to brown in the southern parts of its range in summer. Large densely furred feet offer protection against the snow and stout claws on the forefeet allow it to dig through crusty snow to reach vegetation. Protruding teeth help it scrape moss and lichens from rocks. This species mates in the spring, and a litter of two to eight brown young is born in the summer. The species often congregates in large groups of a hundred or more to feed or rest. A closely related species is the Alaskan Hare (*L. othus*).

Black-tailed Jackrabbit

Lepus californicus L 22 in (56 cm)

Early settlers in the Southwest named this long-eared, lean species "jackass rabbit," later shortened to jackrabbit, although it is clearly a hare in body type, development, and behavior.

KEY FACTS

Back is grizzled buff; chest orangish; underside white; long, black-tipped ears; tail black above, white below.

+ habitat: Brush, pastures, cropland, prairies, and desert

+ range: Western and central U.S., except northern Rockies and plains

+ food: Grasses, forbs, shrubs, cacti, and crops

The Black-tailed Jackrabbit adapts to varied environments ranging from desert scrublands to prairies. Like other hares, the Black-tail has longer legs and larger ears than rabbits, and its ears swivel independently, enhancing its hearing. Deliberate decimation of predators in the Southwest triggered population explosions of jackrabbits, which led to the devastation of crops. Fifteen jacks can eat as much vegetation as one sheep. As in other hares, young are precocial, born furred and mobile, with eyes open. The similar and more northern White-tailed Jackrabbit (*L. townsendii*) has smaller ears and an all-white tail.

|||||||||||||||||||||||||||||||||||||||||||||||||||||||||||||||||||||||||||||||||||||

European Hare

Lepus europaeus or *Lepus capensis* L 27.5 in (70 cm)

This Old World hare, with its distinctive kinked fur, is in decline and gone from parts of its former introduced range. Its antics during breeding season inspired the image of the March Hare.

KEY FACTS

Body is grizzled yellowish brown above; underside is white; large ears are tipped in black.

+ **habitat:** Fields, meadows, and pastures

+ **range:** Southeastern Ontario and adjacent areas of New York; also likely along Hudson River

+ **food:** Grasses, forbs, twigs, and bark

The European Hare is built for speed. Its long hind legs propel it downhill, uphill, or on the straightaway, where it can clock 30 miles (48 km) an hour. It often runs in a zigzag pattern, doubling back on pursuers such as foxes, Bobcats, and hawks. The largest hare in its range, it prefers open spaces where it can hunker down in vegetation, emerging to feed at dusk and dawn. It is active year-round; a female may produce her first litter when snow is still on the ground. Her leverets, or young, disperse to multiple nests for safety. Nursing them in turn, she announces her arrival with a soft grunt, which the leverets answer.

Alaskan Hare/Tundra Hare

Lepus othus L 25.5 in (65 cm)

The Alaskan Hare is similar to the Arctic Hare in appearance, habitat, and adaptation to a cold climate, but the two species' ranges are distantly separated from each other.

KEY FACTS

Body in summer is grayish or reddish brown, white in winter; underside is white; ears are black-tipped.

+ habitat: Tundra, especially near alder thickets and wet meadows

+ range: Coastal and peninsular western Alaska

+ food: Shrubs, especially willow; leaves, berries, and bark

The Alaskan Hare deals with the extreme conditions of its demanding habitat by often ignoring the elements and not taking cover in rain or snow. Reproduction usually is limited to one litter a year, and the precocial leverets are nursed for several months to maximize their size and fortify them for the rigors of winter. But nests are often mere depressions without any lining, and the leverets stay in place and endure the elements as well. The species begins to change to winter white in mid-September. Seasonal camouflage hinders but does not prevent predation by such species as hawks, foxes, weasels, and wolves.

Swamp Rabbit
Sylvilagus aquaticus L 21 in (53 cm)

A more aquatic species than the Marsh Rabbit *(S. palustris)*, the Swamp Rabbit favors a variety of watery environments across its widespread southern range.

KEY FACTS

Body is grizzled orange-brown above, with paler sides; underside is white; eye ring is orange; tail is small.

+ habitat: Bottom-lands, swamps, and marshes

+ range: South-central U.S. from Oklahoma and Texas to northwestern South Carolina

+ food: Grasses, forbs, buds, and bark

The Swamp Rabbit is larger and longer eared than the Marsh Rabbit, although some authorities consider them a single species. Swamp Rabbits fashion surface nests that often are situated at the base of stumps or posts or in holes in logs and stumps. Lined with vegetation and fur, they shelter a litter of two to three young. Unlike most other cottontails, they are born with fur and with closed eyes that open in two or three days. Adult males are territorial and mark their turf with a substance from glands on their chin. The species shares some parts of its range and habitats with smaller Eastern Cottontails *(S. floridanus).*

Desert Cottontail
Sylvilagus audubonii L 14–16 in (36–41 cm)

Also known as Audubon's Cottontail, the Desert Cottontail ranges the arid plains and valleys of the western U.S. states that it calls home.

KEY FACTS

Body is grizzled yellowish brown above; underside is white; ears are long and black-tipped.

+ **habitat:** Deserts, grasslands, stream bank brush, and piñon-juniper woodlands

+ **range:** North Dakota south through Texas and west to California

+ **food:** Plants, fruits, berries, and acorns

Slightly smaller than an Eastern Cottontail, this species has longer and pointier ears and a tail that is grayish above and white below. The Desert Cottontail rests during the day in a shallow form, or depression, in the vegetation. It comes out at sunset to feed. Under threat, it takes shelter among rocks, in vegetation, or in a burrow dug by another species, such as a prairie dog. If it must, it will also climb a low tree. In some parts of its range, the Desert Cottontail may breed year-round, and the female may have up to five litters with two to four young each year. Otherwise, the breeding season runs spring through fall.

Brush Rabbit

Sylvilagus bachmani 11–13 in (28–33 cm)

More social than a number of other rabbit species, the Brush Rabbit tolerates the company of others but maintains a sense of personal space—a zone of separation of at least a foot (0.3 m).

KEY FACTS

Back is dark or gray-ish brown; sides are grayish; underside is whitish; ears are not black-tipped; tail is short.

+ habitat: Brush, brambles, woodland edges, and chaparral

+ range: West Coast from Oregon through California

+ food: Grasses, forbs, berries, and woody plants

The Brush Rabbit returns to its shelter, called a form, between foraging expeditions to groom and rest. Forms are connected to feeding areas by well-traveled runways through thick vegetation. This rabbit is never far from the cover of brush. It is cautious when entering an area, often remaining motionless while it scans for possible danger. When threatened, it thumps the ground with a hind foot and may let out a squealing distress call. It will climb a bush or low tree for safety. Females raise their young, born in litters of two to five up to four times a year in the forms, which they cover with grass before leaving to forage.

Eastern Cottontail
Sylvilagus floridanus L 15–18 in (38–45 cm)

A very common sight in its large range and varied habitats, the Eastern Cottontail displays the powder-puff rear appendage that gives it its common name.

KEY FACTS

Back is grizzled brown with orange nape; sides are paler; underside is white.

+ **habitat:** Thickets, fields, prairies, forest edges, swamps, suburban and urban areas

+ **range:** Central and eastern U.S. to southern Canada

+ **food:** Grasses, clovers, other plants; bark, twigs, and crops

The Eastern Cottontail fits the quintessential image of a rabbit, showing characteristics that separate rabbits from hares: a stockier body, shorter ears and legs, and typically naked, helpless newborns. A female may have up to seven litters a year, averaging five to seven young. If all her offspring and all of their descendants survived and reproduced similarly, this female would be responsible for more than 2.5 billion bunnies in five years—that is, if cottontails' lives were not so short. The New England Cottontail *(S. transitionalis),* which inspired the character Peter Cottontail, has struggled in its range due to habitat loss.

Mountain Cottontail
Sylvilagus nuttallii L 12-14 in (30–36 cm)

This species has taken the cottontail life to a higher level in the
mountain West. Mountain Cottontails are smaller than eastern
species, with proportionately shorter ears and very furry feet.

KEY FACTS

**Body is grizzled buff
with paler gray sides;
underside is white;
white-edged furred
ears.**

+ habitat: Thickets,
sagebrush, conifer
stands, cliffs and other
rocky areas

+ range: Western
U.S. from southern
Canada to California
and New Mexico

+ food: Grasses,
sagebrush, and juniper

This usually solitary species is quite the tree climber,
often climbing junipers. When feeding at dawn and
dusk, primarily on grasses in summer and on sagebrush
and juniper at other times, it sticks close to cover, either
in thickets or rock crevices. For resting and nesting,
it uses forms, or depressions in veg-
etation, as well as burrows likely con-
structed by other species. Nest forms
are lined with grasses and fur and covered over with leaves and twigs. Females
give birth to four to six young from late spring
to early summer, producing up to five litters a year. This species is also known as
Nuttall's Cottontail.

Marsh Rabbit

Sylvilagus palustris L 15 in (38 cm)

At any sign of danger, this rabbit may take to the water, using strong swimming skills to elude predators. It also floats motionless among vegetation, with only eyes and nose above water.

KEY FACTS

Body is grizzled reddish brown; ears are naked inside; tail is very small and grayish; feet have long claws.

+ habitat: Marshes, lakeshores, and other wet areas

+ range: Coastal southeastern U.S. from Virginia to Alabama

+ food: Roots, tubers, and marsh vegetation

Most rabbits swim when pressed, but the Marsh Rabbit does so regularly; it has been seen going at a good clip more than 700 yards (640 m) from shore. The species has shorter legs than most other rabbits and frequently walks. It also stands and walks comfortably on its hind legs. This species lives and hides amid dense marsh and swamp vegetation, following trampled paths. It builds nests, known as forms, among the undergrowth that are covered over with vegetation to protect the young, which are born with fur, unusually for rabbits, but with closed eyes. Up to seven litters of two to four young may be born in a year.

Northern Short-tailed Shrew

Blarina brevicauda L 5 in (13 cm)

The ability to deliver a poisonous bite distinguishes the Northern Short-tailed Shrew, one of the continent's most common shrew species, from other North American mammals.

KEY FACTS

Body is silvery to dark gray; underside is silvery; ears are hidden by fur.

+ habitat: Woodlands, fields, bogs, marshes, and pond and stream banks

+ range: North-central and northeastern U.S. and southern Canada

+ food: Insects, worms, snails, other invertebrates, and small mammals

Despite a tail that measures only an inch (2.5 cm), the Short-tailed Shrew is the largest shrew in North America in body length and weight—weighing up to a hefty 1 ounce (28 g). The species inflicts its poisonous bite on prey; it is often fatal to insects and paralyzing to larger animals such as moles and voles. Caches of immobilized prey appear to be larders in the shrew's underground burrow. Burrows also are fashioned under leaf litter or snow. The female bears litters of three to seven young in burrow nests lined with grass or fur. The offspring can mate in as few as seven weeks and often live less than a year.

Least Shrew
Cryptotis parva L 3 in (8 cm)

Unlike the solitary individuals of many shrew species, the Least Shrew is gregarious, often living communally and sharing a nest with other adults.

KEY FACTS

Back is brown to black, grayer in winter; underside grayish; short, bicolor tail has tufted tip; ears are short.

+ habitat: Fields, brush, woods, marshes, and dry areas

+ range: Central and eastern U.S. to southern Ontario

+ food: Insects, small lizards and frogs, and carrion

One of the smallest North American shrews, the Least Shrew can weigh as little as a nickel. A hyperactive nature and cooperative behavior help it accomplish large tasks, such as burrow construction. Sometimes one adult will tunnel to create a new space while another shifts and packs in loosened soil. Least Shrews seem to communicate with each other by a variety of sounds, including clicking noises. The species is active day and night throughout the year and may store insects it has killed in a tunnel for future use. In southern parts of the range, females may breed year-round; a typical litter is two to seven pups.

Arctic Shrew
Sorex arcticus L 5 in (13 cm)

A tricolor body distinguishes the long-tailed Arctic Shrew.
This species usually is found in open habitats within the boreal
forests of its northern range.

KEY FACTS

Back is dark brown;
sides are paler; under-
side is whitish; bicolor
tail is dark above,
lighter below.

+ habitat: Edges of
marshes and swamps
and forest clearings

+ range: Much of
Canada and into north-
ern plains of U.S.

+ food: Insects and
insect larvae

Shrew species in general are noted for the frantic feeding required to meet high metabolic needs. Among the
group, the Arctic Shrew may feed even more voraciously,
slowing down for only short periods in its quest for food
by day or night. Good eyesight seems to play a part in
its daytime hunting. It has been observed in the chill air
of early morning ambushing torpid
grasshoppers from above, securing
them with jaws and feet, and then
devouring all but the legs and wings.
This species suffers high mortality,
and young that are born in litters of five
to nine do not make it to maturity about
80 percent of the time.

Marsh Shrew/Pacific Water Shrew

Sorex bendirii L 6.5 in (16.5 cm)

North America's largest long-tailed shrew, the Marsh Shrew shares a similar semiaquatic lifestyle with the Northern Water Shrew, although its range is much more limited.

KEY FACTS

Body is blackish brown above; usually dark on underside; tail is usually all dark.

+ **habitat:** Streams, marshes, beaches, and moist woodlands

+ **range:** Northern California to southern British Columbia

+ **food:** Larvae of aquatic insects, earthworms, sow bugs, and other invertebrates

The Marsh Shrew appears entirely at home in watery habitats, having the ability to swim, dive, scull on the surface, and even dash across the water for three to five seconds. Its dark fur appears silvery when submerged due to air trapped within it. The animal uses its long, sensitive, whiskered snout to hunt for aquatic prey. In winter, the Marsh Shrew may be found farther away from a source of water and it eats a wide range of terrestrial invertebrates as well. This species is larger and lacks the more distinctly hair-fringed feet of the similar Northern Water Shrew (*S. palustris*). It is also known as Bendire's Shrew.

|||||||||||||||||||||||||||||||||||||||||||||||||||||||||||||||||||||||||||||||||

Masked Shrew

Sorex cinereus L 3.5 in (9 cm)

Who was that Masked Shrew? Actually, the "mask" is not readily apparent on the face of this common and widely distributed long-tailed shrew.

KEY FACTS

Back is brown, some-times with a dark central line and paler sides; underside is paler; bicolor tail has black tip.

+ habitat: Woods, fields, marshes, and swamps

+ range: Alaska, most of Canada, and northern U.S.

+ food: Insects, young mice, amphibians, and carrion

Despite its abundance, the Masked Shrew remains elusive as it often travels through leaf litter or subsurface tunnels it constructs or appropriates from mice and moles. Like other long-tailed shrews, the species prefers moist environments. It makes a nest of leaves and grasses under logs or in cavities, in which the female may raise up to three litters of four to ten young each year. Life is short for this species, as it is for most shrews, lasting only about 15 months. As are many shrews, the Masked Shrew is a highly nervous animal, frequently dying of fright when threatened or startled by thunder or other loud noises.

Rock Shrew/Long-tailed Shrew
Sorex dispar L 5 in (13 cm)

Among the long-tailed shrews, the Rock Shrew has a very long tail, which it likely uses for balance while scurrying among the rocks and boulders of its habitat.

KEY FACTS

Back is dark gray, underside paler; tail is more than 90 percent total length.

+ habitat: Rocky regions, especially at base of forested slopes; also in artificial rock debris

+ range: Eastern U.S. from Tennessee and North Carolina to maritime Canada

+ food: Insects, spiders, and centipedes

True to its common name, the Rock Shrew is a denizen of rocky terrain in the mountainous regions of eastern North America. It lives mostly several feet underground and was long thought to be rare, as it didn't show up in collection traps near the surface. The Rock Shrew's head is adapted to hunting in rocky terrain; a narrow skull and long, narrow teeth allow the animal to pry insects and other invertebrates from cracks and crevices. Contrary to a usual correlation between larger size and higher latitude within a species, Rock Shrews in northern areas tend to be smaller than those in the southern part of the range.

Smoky Shrew
Sorex fumeus L 4.5 in (11 cm)

This medium-size shrew is mainly nocturnal and is active year-round. Like many other shrews, it echolocates by constantly emitting noises as it hunts for food.

KEY FACTS

Back is grayish brown; underside is paler; fur is grayer overall in winter.

+ habitat: Leaf litter, mossy rocks, and decaying logs in deciduous and coniferous woodlands

+ range: Southeastern U.S. into southeastern Canada

+ food: Insects, centipedes, sow bugs, earthworms, and salamanders

The Smoky Shrew often inhabits areas populated by Short-tailed and Masked Shrews, but it is larger than the Masked Shrew, which it resembles, and has paler feet. In addition, the Smoky Shrew has fairly prominent ears. The three species tend to minimize competition by eating different sizes of insects and other invertebrates. The Smoky Shrew does not reproduce in its first year; instead, it overwinters as a subadult. The adult female may have up to three litters of two to eight pups between March and October of its second year, but has a life span of only about 14 months and probably does not survive into another mating season.

Merriam's Shrew

Sorex merriami L 4 in (10 cm)

One of the few long-tailed shrew species that prefer dry habitats, Merriam's Shrew uses the runways and burrows of small rodents, especially those of vole species.

KEY FACTS

Back is grayish brown; sides are paler; underside is whitish; tail is dark on top, light on bottom.

+ habitat: Sagebrush, arid grasslands, woodlands, and shrublands

+ range: Western U.S. to southern British Columbia

+ food: Insects, spiders, and other invertebrates

The elusive Merriam's Shrew appears in isolated western colonies, where it is found at elevations ranging from 500 feet to about 9,000 feet (150–2,700 m). This small shrew has a distinctively domed head and a bicolor tail, and like many other shrew species, it molts twice a year. While Merriam's Shrew eats a varied diet of adult and larval insects as well as spiders, in the summer months it seems to have a marked preference for caterpillars. Owls often prey on the Merriam's Shrew, which other potential predators tend to avoid, perhaps because of the pungent odor produced by reproductive glands on males' flanks.

||||||||||||||||||||||||||||||||||||||||||||||||||||||||||||||||||||||||||||||

Pacific Shrew
Sorex pacificus L 6 in (15 cm)

Occupying a limited range and somewhat specialized habitats, the long-tailed Pacific Shrew manages to exploit a wide range of animal and vegetable food sources.

KEY FACTS

Body is reddish brown to dark brown above, paler orange-brown below; legs are long.

+ habitat: Moist thickets, brush, stream banks, and mossy fallen logs

+ range: Coastal northern California into Oregon

+ food: Insects, other invertebrates, amphibians, moss, fungi, and seeds

Its large size and reddish color distinguish the Pacific Shrew, the largest of the *Sorex*, or long-tailed shrew species. This species often deals with an abundance of food-procuring opportunities in its habitats by caching the excess it has caught but cannot immediately consume under logs and in similar spots. It first immobilizes insects by sharp, swift bites to the head or thorax. The mostly nocturnal species uses both smell and hearing in hunting and is capable of pulling flying prey out of the air by using sound to locate it. Female Pacific Shrews bear litters containing from two to seven young in the summer months.

Northern Water Shrew/American Water Shrew

Sorex palustris L 6 in (15 cm)

The largest of the long-tailed shrews in eastern North America, the Northern Water Shrew usually lives near running water and is well adapted to water activities.

KEY FACTS

Back is blackish; underside is usually paler; tail has a tufted tip.

+ habitat: Streams, boggy areas in forests, marshes, and pond and lake edges

+ range: Much of Canada, northern U.S. into Sierras, Rockies, and Appalachians

+ food: Insects, spiders, mollusks, and small fish

The Northern Water Shrew feeds in cold, fast-moving streams where its fur traps a layer of air and keeps it buoyant while it swims to capture prey. Its partially webbed hind feet have fringes of stiff hairs that aid swimming and diving and allow it to dash briefly across the water's surface. The shrew's long, slender snout extends from a pair of tiny, beady eyes and is covered with long sensory hairs, or vibrissae, that rotate constantly. The species is active all day year-round, with peaks of activity before sunrise and after sunset. The female may bear three litters a year of two to ten young in a nest of vegetation.

Star-nosed Mole

Condylura cristata L 7 in (18 cm)

A pink, fleshy appendage of 22 radiating tentacles adorns the end of the snout of the Star-nosed Mole—a signature look unique in the mammalian world.

KEY FACTS

Body is dark brown to black; long, hairy tail is thick at base.

+ habitat: Moist areas in woodlands, meadows, and swamps; sometimes drier areas

+ range: Eastern U.S., west to Minnesota and south to Georgia; southeastern Canada

+ food: Aquatic insects, worms, crustaceans, and fish

Looks aren't the only distinguishing feature of the Star-nosed Mole: It leads a much more aquatic life than other moles. Its tunnels often head right through a stream bank and into the water. This mole is an excellent diver and swimmer, paddling with its wide forefeet, and takes much of its prey in muddy stream bottoms. Its nose tentacles probably aid in prey detection as well as manipulation. This species also lives more communally than other moles. Males and females remain together after mating until the young are born in a nest above the waterline. The two to seven young in a litter display nose stars at birth.

Hairy-tailed Mole

Parascalops breweri L 6.5 in (16.5 cm)

A very hairy, short tail and a preference for higher-altitude living help to distinguish Hairy-tailed Moles from other eastern mole species.

KEY FACTS

Back is grayish black; underside is silvery or spotted white; tail is short and bristled.

+ habitat: Well-drained fields, woodlands, and roadsides up to about 3,000 feet (900 m)

+ range: Northeastern U.S. to southeastern Canada

+ food: Worms, insects, other invertebrates, and plants

The Hairy-tailed Mole shows less of a presence on the surface of the ground than the Eastern Mole does. Its travel and feeding tunnels are shallower and not very noticeable when covered with vegetation. Hairy-tails forage on the surface at night. In cold weather, they build deeper burrows, throwing up short molehills on the surface. Individuals are solitary, and the sexes usually come together only for mating. Once a year, the female bears a litter of four to five whitish, wrinkled young, naked except for some hairs on the face and short sensory hairs. If avoided by predators, the species, like other moles, can live three or four years.

III

Eastern Mole/Common Mole

Scalopus aquaticus L 7 in (18 cm)

Bumpy tunnels that push up lawn turf signal a mole's presence, such as that of the Eastern Mole. The shallow tunnels serve as subways and places to rest and feed.

KEY FACTS

In north, back is gray-ish brown with paler underside; smaller and paler in south.

+ habitat: Fields, woodlands, gardens, and lawns

+ range: Eastern half of U.S., except in mountains, and to southern Canada

+ food: Earthworms, insects, other invertebrates; also plant matter

The abundant Eastern Mole demonstrates the basic mole features that make the animal a tunneling machine: strong, twisted forefeet with flattened claws, a body that tapers at both ends, and fur that moves easily backward and forward. Moles use their distinctive forefeet to excavate soil and push it out of the way. In addition to subway tunnels, moles dig deep tunnels that contain a nest chamber, where females raise litters of two to five young. The nest is advertised by the presence of a molehill of surplus soil on the surface. Despite the *aquaticus* in the scientific name, Eastern Moles are not very aquatic.

Townsend's Mole

Scapanus townsendii L 8 in (20 cm)

North America's largest mole, Townsend's Mole makes
a substantial environmental impact, with its deep tunnels,
high molehills, and large nest chambers.

KEY FACTS

It has a dark brown
or black body; short,
largely hairless tail;
and very narrow
snout.

+ **habitat:** Moist
meadows, floodplains,
woodlands, and alpine
meadows

+ **range:** Northern
California to southern
British Columbia

+ **food:** Insects,
worms, other inver-
tebrates, bulbs, and
other plant matter

Townsend's Mole makes elaborate excavations in the
form of shallow tunnels and deeply dug burrows with
secure nest chambers lined with grasses. Unlike other
moles, the Townsend's shows its presence with abundant
molehills, which rise higher than those of other species and
are found in denser concentrations. A single Townsend's
can be responsible for hundreds of hills
in a small area. The soft, velvety fur
of this species made it commercially
valuable for purses, caps, and cloth-
ing trim. Males are somewhat larger
than females, which bear two to
five young in spring after a gestation
period of about a month.

Southern Long-nosed Bat/
Lesser Long-nosed Bat
Leptonycteris yerbabuenae L 3.5 in (9 cm)

These bats practice chiropterophily—"bat love"—a relationship between the mammals and the plants they feed from and pollinate.

KEY FACTS

Body is orangish brown or grayish; face has a small noseleaf; small tail is barely visible.

+ habitat: Deserts, scrublands, and grasslands

+ range: Southern Arizona and south-western New Mexico

+ food: Nectar, pollen, and fruit of agave, saguaro, and other desert plants; sometimes insects

It's all about the flowers for this species of the desert Southwest. At night, the Southern Long-nosed Bat leaves its roosts in caves and mines, and heads for flowering cacti. Homing in on a flower, the bat hovers hummingbird-like and extends its long tongue. Bristles on the tongue increase its surface area and allow a bat to take up large amounts of nectar and pollen. When flowering ends, the bats consume cactus fruit. Noseleafed bats tend to emit a weak sonar signal, giving them the nickname "whispering bats." Females have their young in large maternity colonies. This species, also known as *L. curasoae*, is endangered.

California Leaf-nosed Bat

Macrotus californicus L 3.5 in (9 cm)

The California Leaf-nosed Bat needs no introduction: The combination of noseleaf and enormous ears distinguish it from other North American bats.

KEY FACTS

Body is grayish; nose has distinctive, erect noseleaf; ears are very large and rounded; tail extends beyond membrane.

+ habitat: Desert scrublands

+ range: Southern California, Nevada, and Arizona

+ food: Insects and larvae, including cater-pillars; cacti fruit

Barely able to move on the ground, the California Leaf-nosed Bat is master of the air. Short, broad wings lend it great maneuverability. Large ears—a characteristic of strong fliers—enhance hearing, allowing the bat to detect insects among thick vegetation. This species often hovers over trees and other plants, gleaning prey such as cater-pillars and grasshoppers from vegetation or the ground. The leaf may indicate that echolocation takes place through the nostrils. These bats seek out roosts in mines and large caves, but they do not clus-ter together and seem to avoid the direct touch of other roosting individuals.

Western Mastiff Bat

Eumops perotis L 7 in (18 cm)

Large ears that join across the skull and project over the eyes give the Western Mastiff Bat the wrinkled canine appearance that suggests its common name.

KEY FACTS

Body fur is dark gray-ish brown or brown with white roots that often are visible.

+ **habitat:** Deserts, canyons, scrublands, and urban areas

+ **range:** Southwestern U.S. to northern California

+ **food:** Mainly flying insects, including moths, beetles, and bees

The largest U.S. bat, weighing all of 2 ounces (56 g), the Western Mastiff displays a long tail extending well past the interfemoral membrane that is characteristic of free-tailed bats. Its size and narrow wings make the bat a fast but not agile flier. These bats roost in high places such as crevices in high cliffs, allowing them to drop and launch themselves. They often climb high into the air and utter loud, piercing calls. They mate in spring, and unlike many other bat species, adult males sometimes join maternity colonies. The species, also known as the Western Bonneted Bat, leaves a telltale urine stain on cliff faces.

Brazilian Free-tailed Bat/
Mexican Free-tailed Bat
Tadarida brasiliensis L 4.5 in (11 cm)

A protruding tail explains the species' common name. The species forms the largest communal roosts in the bat world.

KEY FACTS

Back is uniform gray-ish brown; underside is slightly paler; ears are wide.

+ habitat: Scrub-lands, deserts, and other open areas

+ range: Southern U.S., ranging farther north in the West than the East

+ food: Flying insects, including moths

Brazilian Free-tailed Bats live a life of superlatives. Narrow wings help it achieve renown as one of the fastest-flying bats, reaching speeds of up to 60 miles (96 km) an hour. It forms colonies in the millions that rank as the largest assemblages of a mammalian species anywhere in the world. Leaving their roosts in caves and buildings and under bridges at sunset, these bats spiral upward into a column and then head off for a night of hunting. Returning to the roost, they free-fall from the sky at the entrance. The species migrates to Mexico for the winter, enduring a flight of up to 1,000 miles (1,600 km).

Big Brown Bat

Eptesicus fuscus L 4.5 in (11 cm)

Familiar to many, Big Brown Bats circle above trees in evenings and enter attics, barns, and belfries to roost. This species is greatly impacted by white-nose syndrome, a fungal disease.

KEY FACTS

Body fur is dark at the roots, becoming brownish at the surface; broad face and rounded ears are blackish.

+ **habitat:** Woodlands, fields, suburban and urban areas

+ **range:** Throughout U.S. and Canada, except far northern areas

+ **food:** Mainly flying insects, especially beetles

Size and a wide-ranging presence help identify the Big Brown Bat. Its tail extends beyond the membrane between its back legs. The species emerges to hunt after sunset, using echolocation to avoid obstacles and home in on insects. It mates before hibernation, although fertilization is delayed until spring. Females roost in maternity colonies that average 25 to 75 individuals. In the West they tend to have a single offspring and in the East, twins. On warm winter days these bats may awake and forage; a bat seen in midwinter may well be a Big Brown. This species plays a major role in controlling agricultural pests.

Eastern Red Bat

Lasiurus borealis L 4.5 in (11 cm)

Hanging by one leg from a tree branch, the solitary Eastern Red Bat resembles a dead leaf, a fact that may deter predators such as blue jays, opossums, hawks, and owls.

KEY FACTS

Male is orange to reddish brown; female less red and frosted with white; both have white patches on shoulder and thumb.

+ **habitat:** Fields, woodlands, and urban areas

+ **range:** Central and eastern U.S. and Canada (except far north), and southern Florida

+ **food:** Flying insects, especially moths and beetles

Reddish fur, long wings, and a furred tail membrane that can be wrapped around the body like a blanket distinguish the Eastern Red Bat. The males and females show a marked difference in coloring, the males being distinctly redder. These bats usually do not stray far from the roost, often in dense foliage, to hunt insects. Bats in the North may migrate to the southern part of the range in winter.

Although twins are common among many bat species, the Eastern Red female may give birth to three or four young at one time. This species is sometimes found on the ground and may hibernate on the ground among leaf litter.

Hoary Bat

Lasiurus cinereus L 6 in (15 cm)

These bats are called "hoary" for the grayish white tips of their brown fur. The silvery coat helps them blend in with lichens that grow on evergreen trunks.

KEY FACTS

Body is brownish gray tipped with white; face is surrounded by yellow band; ears are short, round, and black-edged.

+ **habitat:** Deciduous and coniferous woodlands near water

+ **range:** Most of U.S. and Canada, except far north

+ **food:** Flying insects, especially moths, and also smaller bats

North America's most widespread bat, the Hoary Bat is more abundant in the western half of its range. This large bat has powerful wings built for long-distance migration, and it often chatters audibly in flight. The species has appeared far out of range and has been spotted in such distant places as Iceland and the Orkney Islands. It is the only wild land mammal to have reached and established itself in the Hawaiian Islands.

This bat migrates in large waves; in spring, the pregnant female moves north ahead of the male and gives birth to two young in May or June. Like all bats, the Hoary is a fastidious groomer.

Northern Yellow Bat
Lasiurus intermedius L 5.5 in (14 cm)

Like the Seminole Bat, the Northern Yellow prefers to roost in clumps of Spanish Moss. It can be distinguished from other yellow bats by its partially furred tail membrane.

KEY FACTS

Fur is long, silky yellowish orange to yellowish brown with a grayish wash; the outer tail membrane is furred.

+ habitat: Forest edges with Spanish Moss and palm groves

+ range: Coastal southeastern U.S.

+ food: Mainly flying insects, including mosquitoes, flies, and bees

Northern Yellow Bats frequently hunt over open areas with water and are seen over coastal golf courses and beaches. Small groups of males and females roost together, and the species may form large feeding aggregations near the end of summer. Females often bear three or four young in summer maternity colonies; in winter, the sexes tend to segregate. The similar Southern Yellow Bat *(L. ega)* is smaller and shows a less grayish wash to its fur. In the United States it is found primarily at the southern tip of Texas. Like other *Lasiurus* species, the Northern Yellow Bat is a vesper bat, a member of the world's largest bat family.

|||||||||||||||||||||||||||||||||||||||||||||||||||||||||||||||||||||||||||||||||||||

Seminole Bat/Mahogany Bat

Lasiurus seminolus L 4 in (10 cm)

Similar in size, wing shape, and color to the Eastern Red Bat, the Seminole Bat also overlaps with the southeastern portion of the Eastern Red's range.

KEY FACTS

Both male and female have reddish brown fur, white-tipped on back; shoulder and thumb have white patches.

+ habitat: Swamps, bogs, and forests, near water

+ range: Southeastern U.S.; isolated populations elsewhere

+ food: Flying insects, including flies and beetles; also ground insects

Unlike the similar Eastern Red Bat, male and female Seminoles share the same fur color. Eastern Red females are usually duller than Seminoles of either sex. Seminoles fly straight and swiftly, emerging after dusk to hunt on the wing or to glean insects from vegetation or the ground. They often roost individually in Spanish Moss or sometimes in pairs in the same clump of the epiphytic plant. Collection of Spanish Moss for commercial purposes has significantly reduced the habitat available to the Seminole Bat. These bats do not enter deep hibernation for the winter, but may emerge from periods of inactivity on warm winter days.

Western Pipistrelle

Pipistrellus hesperus L 3 in (8 cm)

North America's smallest bat, the Western Pipistrelle can weigh as little as a tenth of an ounce (2.8 g). Even so, adult females usually give birth to twins.

KEY FACTS

Body is blondish or yellowish; black mask and ears provide contrast.

+ habitat: Grasslands, scrublands, canyons, and woodlands near water

+ range: Arid parts of the western United States to Washington

+ food: Mainly small swarming insects, including mosquitoes and stoneflies

The tiny Western Pipistrelle has diurnal tendencies, sometimes beginning the nightly hunt before sunset, its slow, fluttering flight suggesting that of a butterfly. After a period of rest, it reemerges and often stays active into the morning in pursuit of swarming insects. Pale coloring echoes the species' customary habitats in arid parts of the western United States. Western Pipistrelles often roost in cracks and crevices of canyons and cliffs or in caves, always near water. Young weighing only about 0.05 ounce (1.4 g) are born in small maternity colonies; within a month or so, they are indistinguishable from adults.

Eastern Pipistrelle

Pipistrellus subflavus L 3.5 in (9 cm)

A tad larger than its western counterpart, the very abundant Eastern Pipistrelle often returns to hibernate in the same spot in the same cave each year.

KEY FACTS

Fur is dark at base, light in middle, and brown on tips; has pinkish brown face and forearms.

+ **habitat:** Farmland and woodlands, near water

+ **range:** Eastern half of U.S. and extreme southeastern Canada

+ **food:** Flying insects, including beetles and flies

Tricolor fur distinguishes this tiny bat, so small that it is sometimes mistaken for a large moth. Primarily a woodland species, the Eastern Pipistrelle often roosts in dense foliage in summer, emerging at dusk to flutter through trees in search of insects. This bat remains in the same general area year-round. It carefully chooses a location in the cave where it overwinters, seeking the right balance of chilly temperatures and moisture to ensure that it does not dry out while hibernating. Females mate in the fall and store the sperm for delayed fertilization, or mate in the spring, giving birth usually to twins.

Rafinesque's Big-eared Bat

Corynorhinus rafinesquii L 4 in (10 cm)

Elaborate, supersize ears are the hallmark of Rafinesque's Big-eared Bat. The tragus, a feature of the outer ear that stands at the entrance to the inner ear, is also prominent in this species.

KEY FACTS

Body is gray to brown on the back and lighter to whitish on the underside; bumps of flesh appear behind the nostrils.

+ **habitat:** Forested areas near water sources

+ **range:** Southeastern United States

+ **food:** Flying insects, especially moths

The large ears and tragus of this bat species likely serve as direction detectors. The tragus stands erect, even when the bat is resting and coils up its ears and tucks them under its wings. When disturbed by a noise, the bat moves its head and waves its ears as though tracking echoes. The elusive Rafinesque forages for insects—often hovering in flight—later in the evening than many other bats, and returns to its roost while it is still dark.

This species was partial to roost-ing in the "twilight zone" at cave entrances, but it now roosts in buildings and other structures as human activity has breached these habitats.

Pallid Bat

Antrozous pallidus L 4.5 in (11 cm)

The Pallid Bat pursues a unique bat lifestyle: It feeds almost entirely on items from the ground, taking such prey as crickets, grasshoppers, scorpions, and lizards.

KEY FACTS

Body is sand colored; ears are enormous and point forward; nose is piglike.

+ habitat: Deserts, grasslands, canyons, and mixed forests

+ range: Southwestern U.S. to southern British Columbia

+ food: Ground and flying insects, scorpions, and nectar

The Pallid Bat is colored to match its desert surroundings. Like other big-eared bats, this species can hover in flight. But it lands often, gleaning vegetation and the ground for insects. Because of prickly prey and prickly vegetation, the Pallid Bat frequently displays holes in its wing and tail membranes. The Pallid Bat also visits cactus flowers, serving as an important pollinator. It roosts in rock crevices and buildings. During a night of foraging, it may come to rest in open, sheltered spots such as bridges and porches. This noisy bat calls to others and is known to bare its teeth and buzz when frightened or angered.

Silver-haired Bat

Lasionycteris noctivagans L 4 in (10 cm)

The Silver-haired Bat's silver-frosted fur gives it a distinctive appearance. The bottom half of the interfemoral membrane between its back legs also is furred.

KEY FACTS

Back is dark brown to black with silver frosting; face is dark; ears are rounded.

+ **habitat:** Grasslands, scrublands, and woodlands near water

+ **range:** Most of U.S. and Canada, except far northern areas

+ **food:** Mainly small flying insects including beetles, flies, and moths

Soon after sunset the Silver-haired Bat begins to hunt, often flying low and slowly in search of flying insects, especially its preferred prey: moths. It seldom is found far from water, roosting solitarily or in small groups in mines, caves, hollow trees, rock crevices, houses, and under bark. Silver-hairs mate in autumn. As with other bat species, fertilization is delayed until spring, when females form much smaller maternity groups than other species do. After a two-month gestation, they give birth, usually to twins. Some Silver-hairs appear to migrate; others stay put and may enter a state of torpor in winter.

Long-eared Myotis
Myotis evotis L 3.5 in (9 cm)

Long, dark ears—the better to hear insects rustling in the vegetation—characterize the Long-eared Myotis. This bat often appears to turn off echolocation when hunting.

KEY FACTS

Back color ranges from yellowish to dark brown; face and ears are black.

+ **habitat:** Woodlands, shrublands, grasslands, and agricultural areas

+ **range:** Temperate areas of western U.S. and Canada, to about 9,200 feet (2,800 m)

+ **food:** Flying insects, including moths and beetles

Long-eared Myotis leave their customary roosts in rocky outcroppings, dead trees, caverns, and buildings to glean insects from dense vegetation. To do so, they may hover momentarily to capture their prey. They also hunt among grass and trees along bodies of water. Active a bit longer into the night than other bats, they may hunt closer to the ground as the night air gets cooler. Females form small maternity colonies, often with males roosting nearby. They bear a single offspring in the summer months. Long-lived for a small mammal, this species can survive up to 22 years, although a typical life span is much shorter.

Little Brown Bat/Little Brown Myotis
Myotis lucifugus L 3.5 in (9 cm)

The Little Brown Bat is one of North America's most abundant bat species. As natural sites disappear, this highly adaptable bat often establishes colonies in houses and other buildings.

KEY FACTS

Back varies from tan to reddish to dark brown; underside is buff or grayish white.

+ habitat: Forested areas, often near water

+ range: U.S. and Canada, except far north and southwestern and south-central U.S.

+ food: Aquatic insect larvae, flies, moths, beetles, and other insects

The Little Brown Bat is a basic, no-frills species—small, glossy-furred, and lacking any facial ornamentation such as noseleafs or chin leafs. These bats mate in the fall at their hibernation sites, often in caves, but fertilization occurs with spring emergence. Females form maternity colonies with up to thousands of individuals in warm places, such as attics. When a single offspring is born after 50 to 60 days, the mother receives it in the membrane between her legs, which is cupped like a basket. Within a month it will attain adult weight: all of a quarter of an ounce (7 g). These little bats can live as long as 40 years.

Ocelot

Leopardus pardalis L 37–50 in (94–127 cm)

The beautiful, endangered Ocelot is a species that generally keeps to itself. Far less is known about its habits than about those of other wild cat species in North America.

KEY FACTS

Head, back, limbs, and tail have black spots and markings on grayish, reddish, or buff background.

+ **habitat:** Forests and brushland

+ **range:** Extreme southern Texas; former U.S. range much larger

+ **food:** Rabbits, other small mammals, birds, reptiles, fish, frogs, and some invertebrates

Fast on the ground and adept at tree climbing if pursued, the Ocelot typically rests in foliage by day and hunts by night. Both male and female Ocelots are territorial, and males try to claim females whose territories overlap with their own. The species' distinctively marked coat, which helps it blend with the dappled sun and shade of its environment, has made it a target for its fur and for the pet trade. Its endangered status makes those activities illegal in the United States and in some Latin American countries in its range. The tiny population surviving in Texas faces habitat loss and the perils of living near roads.

Canada Lynx

Lynx canadensis L 27–42 in (69–107 cm)

This stealthy cat of the boreal forest creeps up on its big cat feet, stalking undetected to within a few bounds before pouncing on its prey—usually the Snowshoe Hare.

KEY FACTS

Back is grayish buff mottled with brown; black-tipped tail is bobbed; ears have black tufts; face has sideburns.

+ **habitat:** Dense vegetation in northern boreal forests

+ **range:** Alaska and much of Canada, northwestern U.S., upper Midwest, and upper Northeast

+ **food:** Primarily Snowshoe Hares

The fate and well-being of the Canada Lynx is tied closely to that of the Snowshoe Hare. When hare populations crash, so do the lynx's, almost simultaneously. Breeding takes place in winter, even in famine, but the typical litter of one to four kittens—born in a den after 60 to 65 days—may suffer high mortality. A robust lynx litter looks and sounds much like a litter of domestic cats, but larger and a lot louder. Lynxes are larger than Bobcats, their ear tufts and sideburns are longer, and their tails are entirely black-tipped. The Canada Lynx's luxurious fur has made it a prime target of trappers for hundreds of years.

Bobcat

Lynx rufus L 24–47 in (61–119 cm)

The Bobcat's bobbed tail offers no advantage in balance or navigation, but with its white underside, it acts as a location flag for kittens trailing their mother in dense underbrush.

KEY FACTS

Back varies from brown to grayish with spots, even to solid black; underside is white with black spots; tail tip is black.

✦ habitat: Forests, swamps, grasslands, and mountains

✦ range: U.S. and southern Canada; absent in parts of Midwest and East

✦ food: Rabbits, hares, rodents, birds, and deer

This solitary hunter, the most widespread wild cat in North America, is about twice the size of a domestic cat. It mainly hunts rabbits and hares but is capable of killing healthy deer. Bobcats tend to be nocturnal, although they can be spotted during the day. Adults come together at breeding time, but severe weather may find some sharing a rock shelter without much interaction. The one to six young born in spring or summer stay with their mother and disperse before the next litter arrives. The species adapts well to disturbed habitats, but heavy trapping has greatly reduced its numbers or eliminated it in some areas.

Mountain Lion/Cougar/Puma

Puma concolor L 5–8 ft (1.5–2.4 m)

The Mountain Lion, largest of the cats in the United States and Canada, answers to many regional names. In addition to Cougar and Puma, it is called Panther, Catamount, and Painter.

KEY FACTS

Body is tan or russet with a whitish underside; long tail has a black tip.

+ **habitat:** Desert, forest, and mountain wilderness areas

+ **range:** Western United States into British Columbia and Alberta and southern Florida

+ **food:** Primarily large hoofed mammals

The Mountain Lion has certain traits that make it more similar to a house cat than to a lion. The species is a purring cat, not a roaring one, and it leads a solitary life except at mating time. It also employs a stalk-and-ambush hunting strategy. Cougars may mate at any time of year. A litter of one to six spotted kittens is born in a secure den after a 90-day gestation; their spots will soon give way to the single color reflected in the species name: *concolor,* or uniform color. This large cat once ranged across North America, but now is confined mainly to the West. The Florida Panther, a highly secretive subspecies, is endangered.

Coyote

Canis latrans L 3.5–4 ft (1–1.2 m)

As the Gray Wolf retreated in North America, the adaptable and resourceful Coyote fanned out from the prairies and took its place—and has since added even more territory to its range.

KEY FACTS

Back ranges from grayish to tawny brown; muzzle and legs are long; bushy tail is black-tipped.

+ habitat: Deserts, grasslands, woodlands; agricultural, suburban, and urban areas

+ range: Throughout U.S. and Canada, except some northern areas

+ food: Mammals, birds, snakes, insects, and carrion

The Coyote has met few habitats not to its liking. Able to adapt its living, hunting, and breeding styles to many areas, the species has even become an entrenched urbanite. Coyotes often are confused with foxes, which are smaller and have pointier faces, and with some breeds of domestic dogs, which usually have less bushy tails. They sometimes mate with dogs, producing a fertile, though not robust, offspring known as a "coydog." Coyotes live as individuals, mated pairs, or in family-based packs. Pack hunters are able to take down larger prey, such as deer. Eastern Coyotes tend to be larger than western ones.

Gray Wolf
Canis lupus L 4–6 ft (1.2–1.8 m)

The Gray Wolf once had free roam in much of North America, but centuries of trapping, shooting, and poisoning, often for a bounty, have confined it mainly to the northern U.S. and Canada.

KEY FACTS

Color from white to black, most often grayish on back; underside is usually lighter; ears are small; tail is long and bushy.

+ habitat: Forests and tundra in wilderness areas

+ range: Alaska, western and central Canada to northwestern U.S.

+ food: Hoofed mammals, hares, and rodents

The Gray Wolf most often lives in packs; lone wolves are rare. Packs comprise a half dozen or more individuals that live and hunt together to bring down large hoofed mammals. In a pack, there is a dominant, or alpha, male and an alpha female; they fill leadership roles and hold reproduction rights and other perks, such as the first go at a kill. Other adult wolves in the pack help feed and look after the pups of the alpha couple. A litter averages six young, born in spring through early summer. The issue of state and federal protection of the Gray Wolf as an endangered species continues to spark heated debate.

Red Wolf

Canis lupus rufus or *Canis lupus* L 4–5 ft (1.2–1.5 m)

This smaller, slimmer, and redder relative of the Gray Wolf once populated much of the southeastern United States. Last-minute efforts in the 1970s saved it from extinction.

KEY FACTS

Back is gray to black to reddish brown; sides are tan; underside is white.

+ habitat: Formerly in forests, bottomlands, grasslands, other dense vegetation

+ range: Reintroduced to coastal North Carolina and South Carolina, Tennessee, and Gulf islands

+ food: Rodents, rabbits, deer, other mammals, and birds

The Red Wolf is similar in appearance to the Coyote, but the latter has more gray on its muzzle, which usually is pointier than that of the Red Wolf. The spread of the Coyote displaced the Red Wolf—an animal that lives in mated pairs or small packs—which already was reduced to low levels from targeted elimination and environmental disruption. Coyotes interbred with the wolves, leaving only a tiny Red Wolf population on the border of coastal Texas and Louisiana. Wolves there were evacuated for captive breeding and founded a population that has allowed small, but successful, reintroductions in the wild, starting in 1987.

Gray Fox

Urocyon cinereoargenteus L 2.5–3.5 ft (0.8–1 m)

The Gray Fox often uses its sharp, curved claws to scramble catlike up a tree to escape predators, to rest, or to forage. It can also jump nimbly from branch to branch.

KEY FACTS

Body is grizzled gray with reddish ears, neck, and sides; underside is white; bushy, tail tip is black.

+ **habitat:** Woodlands, old fields, brushlands, and rocky areas

+ **range:** Much of U.S., except Northwest; parts of southern Canada

+ **food:** Mammals, lizards, frogs, insects, nuts, fruit, and carrion

The Gray Fox is active at dawn, dusk, and during the night, availing itself of a wide range of animal and vegetable food sources. During the day it rests, often in dense vegetation or in the shade of a tree or rock. Dens mostly come into play during whelping season. The female gives birth to an average of four kits in a hollow log, rock crevice, abandoned building, brush pile, or burrow abandoned by other species, including the Red Fox and Groundhog. The male fox, or dog, helps care for the young, but the female fox, or vixen, meets most of their needs until they can fend for themselves at about seven months.

Arctic Fox

Vulpes lagopus L 2.5–3.5 ft (0.8–1 m)

The grayish brown coat of the Arctic Fox molts to a snow-compatible white one as winter begins. Other Arctic adaptations include a compact body, short legs, small ears, and furred feet.

KEY FACTS

Body is grayish brown above and cream on the underside in summer; coat turns white in winter.

+ habitat: Tundra, coastal areas, and ice floes

+ range: Arctic areas of Alaska and Canada

+ food: Rodents, hares, fish, birds, eggs, carrion, and garbage

The Arctic Fox needs camouflage because it remains active all winter, retreating to a snow bank or den only in severe weather. A rare variety of the species has a coat of bluish gray in summer that gets paler in the winter. Arctic Fox mates work together to rear the 6 to 12 pups born in a den that is reused and enlarged for decades or longer, because of the difficulty of new excavation in the permafrost. An old den may be a honeycomb of tunnels and entrances; the fox often sits watch on top. The Arctic Fox eats a wide range of seasonal foods, including the lemming, its traditional prey. It also scavenges Polar Bear kills on ice.

Swift Fox

Vulpes velox L 2.5–3 ft (0.8–0.9 m)

True to its name, the lithe Swift Fox, with its elegant, wide-set ears and dark-sided muzzle, can outrun all but the fastest predators on its home turf of flat, open terrain.

KEY FACTS

Back is grizzled gray with yellowish cast; neck, sides, and legs are yellowish orange; throat and underside are white.

+ **habitat:** Grasslands, shrublands, and desert

+ **range:** High plains from Texas into southern Canada

+ **food:** Rodents, rabbits, birds, lizards, insects, and vegetation

The Swift Fox is well camouflaged to blend with tawny grasses of the high plains; it sometimes rests in the sun outside its underground den. This fox appears to have strong family bonds. Two to seven young are born in the den as early as February; they emerge at three weeks, and the parents begin training them to hunt. If the female dies, the male may raise the pups alone. In the fall, they are ready to go off on their own. A hundred or so years ago, the Swift Fox population declined drastically from habitat destruction because of conversion of prairie to cropland, trapping, and poison intended for other species.

Red Fox

Vulpes vulpes L 3–3.5 ft (0.9–1 m)

Settlement and development, along with the disappearance of wolves, helped the Red Fox expand its horizons and overtake the Gray Fox in abundance in eastern woodlands.

KEY FACTS

Body is usually orangish red; legs and ears are black; tail tip is white.

+ **habitat:** Fields, forests, brushland, marshes, suburban and urban areas

+ **range:** Most of U.S. and Canada, except parts of western U.S.

+ **food:** Mammals, birds, eggs, frogs, insects, fruit, berries, and carrion

European settlers found the Gray Fox to be frustratingly arboreal, so they imported the Red Fox from Europe for foxhunting, only to discover that the species was a native here. The Red Fox hunts in a catlike manner, sneaking up to pounce on prey with a pronounced jump. The species seems to mate for life. The male and female dig a den in winter or find a suitable cave or crevice in which the female gives birth to about five kits in the spring. Fox families often incorporate daughters from the previous year. From a distance, foxes are sometimes confused with domestic cats; larger ears and brushy tails help identify the fox.

American Black Bear

Ursus americanus L 4-6 ft (1.2-1.8 ft)

Known for shaking down visitors to national parks and trashing cars, tents, and cabins, the American Black Bear achieved an eclectic appetite through exposure to the human world.

KEY FACTS

Color black to reddish to white; dark bears have lighter snouts.

+ habitat: Forests, swamps, tundra, and mountains

+ range: Canada and western U.S.; north-eastern U.S.; isolated populations elsewhere

+ food: Grasses, acorns, fruits, insects, mammals, fish, birds, and carrion

More adaptable than the other two North American bear species (*U. arctos* and *U. maritimus)*, the American Black Bear also is the most wide-ranging. It has larger ears and eyes and a longer snout than the other species. Males are larger than females, weighing up to 900 pounds (400 kg). Black Bears climb trees efficiently and can sprint up to 35 miles (56 km) an hour. They den in winter in such spots as caves, under fallen trees, and in tree hollows. They do not hibernate—or undergo a marked metabolic slowdown—but are merely dormant. Females give birth to one to three cubs in the den and nurse them throughout the winter.

Grizzly Bear/North American Brown Bear

Ursus arctos horribilis L 5.5–8.5 ft (1.7–2.6 m)

In some areas these bears have brown fur with grayish tips that give the Grizzly Bear its common name: Grizzled means "grayish."

KEY FACTS

Body is tan, grayish, reddish, or brown; ears are small, rounded; shoulder is humped.

+ habitat: Forests and open area in mountains; rivers and coasts

+ range: Northern U.S. Rockies into western Canada and Alaska

+ food: Vegetation, nuts, berries, insects, birds, eggs, fish, mammals, and carrion

A Grizzly Bear in profile with all feet on the ground shows a higher shoulder; a Black Bear shows a higher rump. The Grizzly's large forefeet armed with up to 4-inch (10 cm) curved claws can deliver a lethal blow. Adult males dominate females and young bears, especially during the feeding frenzy of the salmon run. Mothers will secure their offspring, born in the winter den, from adult males, which are known to sometimes kill and eat the young. Like Black Bears, Grizzlies take advantage of food sources associated with human activity, especially in national parks, although measures are taken to mitigate this situation.

Kodiak Bear

Ursus arctos middendorffi L to 9 ft (2.7 m)

The largest North American land mammal, outweighing even the Polar Bear, the Kodiak Bear is a subspecies of North American Brown Bear. It is named for Alaska's Kodiak Island.

KEY FACTS

Body is blondish brown to dark brown; has pronounced shoulder hump.

+ **habitat:** Grasslands, meadows, wet tundra, shrublands, and spruce forest

+ **range:** Kodiak Island and other islands and coast of southern Alaska

+ **food:** Vegetation, nuts, berries, fish, small mammals, and carrion

Kodiaks of southern Alaska display the same humped shoulder, scooped face, and long claws that other Brown Bears do, including inland Grizzlies. Yet these beachcombers can weigh up to 1,600 pounds (700 kg), more than an average Grizzly. Like other Brown Bears, Kodiaks feed to maximize weight gain, utilizing a variety of foods to create the bulk that will carry them through many months in the den without eating and drinking—or urinating or defecating. Alaska protects its Kodiak Bears, allowing very regulated hunting and safeguarding traditional habitat. Aggressive encounters between Kodiaks and humans tend to be rare.

Polar Bear
Ursus maritimus L 6–8.5 ft (1.8–2.6 m)

The massive Polar Bear spends much of its time adrift on pack ice and floes, stalking the Ringed Seals that compose its main source of food.

KEY FACTS

The fur is white to yellowish white; the neck is long; the head is small; the eyes and nose are black.

+ habitat: Pack ice and ice floes, rocky shores, and islands

+ range: Canadian Arctic coasts and islands

+ food: Seals, walrus, small whales, fish, ducks, plants, and carrion

The Polar Bear has developed a unique Arctic lifestyle. This strong swimmer has an insulating white coat that provides effective camouflage. After mating, the adult female digs a den in a snowbank that is vented to the surface. The female enters the den in the fall and gives birth, usually to twins, within a few months; mother and cubs emerge in the spring. Like other bear species, the Polar Bear seeks out food associated with human habitation, including garbage dumps. Attempts to relocate "nuisance" bears often fail, as they are accustomed to traveling hundreds of miles during the year in search of food and then returning.

North American River Otter

Lontra canadensis L 38–49 in (97–124 cm)

Otters often build play into their daily routines. For example, family groups body-sled one at a time down snowbanks or slippery mud banks, often ending the run with a loud splash into a river.

KEY FACTS

Body is brown above, silvery below; short legs have webbed feet; tail is thick at base.

+ habitat: Streams, rivers, lakes, swamps, and coastal areas

+ range: Alaska and Canada, northwestern and eastern U.S.; reintroduced elsewhere

+ food: Fish, crayfish, frogs, small mammals, and insects

Among the North American River Otter's aquatic adaptations are a streamlined body, waterproof coat, and flaps of skin that close nose and ears when it dives. Otters prefer bodies of water replete with dens and burrows of species such as Muskrats and American Beavers. Otters mate in water, and implantation is often delayed as long as nine months. Females bear one to six young in a den or lodge; males do not participate in rearing the young. Overhunting and pollution decimated otter populations, which once occurred in much of North America. The species is now making a comeback through protection and reintroduction.

Wolverine
Gulo gulo L 32–46 in (81–117 cm)

The Wolverine's fierce reputation is not hype: The strength of this giant weasel likely exceeds that of any other mammal its size. It can take down large hoofed mammals.

KEY FACTS

Body brown to black; lighter face rim and ears; often has white chest patches.

+ **habitat:** Boreal forests, tundra, taiga; mountainous areas

+ **range:** Northern Canada to northern Rockies; some in western coastal states

+ **food:** Mammals, birds, berries and other plant material, and carrion

The elusive Wolverine is equipped for long, northern winters. It does not migrate or hibernate, but continues its opportunistic search for food with the aid of a thick fur coat and broad, furred feet that allow it to lope over crusty snow. Its signature weasel scent glands exude a highly odoriferous substance that marks territory and stakes a claim on its kills. Strong, smart, and often at odds with human activities, the species was targeted for extermination with traps and poisons, and suffered widespread habitat destruction. These factors nearly wiped out the Wolverine in the United States; only remnant populations survive.

American Marten/Pinze Marten

Martes americana L 18–24 in (46–61 cm)

Mature evergreen forests form the prime habitat of this solitary, catlike carnivore. Adult American Martens often establish a home range of up to 15 square miles (39 sq km) when food is scarce.

KEY FACTS

Body is dark brown; head is paler; throat is whitish or orange.

+ **habitat:** Coniferous and mixed forests

+ **range:** Much of Alaska and Canada; small populations elsewhere, mainly in western U.S.

+ **food:** Small mammals, birds, eggs, insects, worms, nuts, berries, and carrion

A skilled hunter in the trees or on the ground, the American Marten exploits a wide range of food sources during the year. Agile and fast enough to pursue a squirrel from branch to branch—and catch it—the marten can also track a Snowshoe Hare or vole in the snow, traveling when necessary through tunnels made by other animals. This species once ranged well into the United States, but has suffered significant habitat loss; reintroduction programs have been successful. It differs from the similar American Mink by its less uniform coat color and dark appendages and from the Fisher by its smaller, more slender size and shape.

Fisher

Martes pennanti L 29–47 in (74–119 cm)

Despite the name, Fishers don't fish. They primarily target hares, rabbits, and small rodents for a meal. They also successfully take on adult porcupines, launching an attack from the front.

KEY FACTS

Body has long dark brown fur; head is grizzled; ears are small; nose is pointy.

+ **habitat:** Mature coniferous and mixed forests

+ **range:** Southeastern Alaska through Canada into eastern U.S. and western mountains

+ **food:** Rodents, rabbits, hares, birds, fruits, nuts, and carrion

Fishers are omnivores that often exploit the same habitats and food resources as American Martens. Where the two occur together, the larger Fisher often takes the larger prey. Foresters welcome the Fisher's willingness to go after North American Porcupines, as the quilled rodents decimate trees by their fondness for inner tree bark. The Fisher bites the porcupine in the face and neck before rolling it over to tear into the belly. A female Fisher mates about a week after giving birth. The embryo doesn't implant for another ten months or so. One to six young are born in the early spring, often in a den in a hollow tree or log.

Ermine/Stoat/Short-tailed Weasel
Mustela erminea L 7–12 in (18–30 cm)

Whether in its winter white or summer brown coat, the Ermine is a long, sleek, and very efficient hunter, able to pursue small mammals into any burrow or hole it can get its head into.

KEY FACTS

Back is brown to reddish in summer (whitish in winter); underside is white; tail tip is black.

+ **habitat:** Forests, tundra, meadows, marshes, riverbanks, and hedgerows

+ **range:** Alaska and Canada into western U.S., upper Midwest, and Northeast

+ **food:** Rodents, rabbits, frogs, and earthworms

Lightning-quick reflexes make the Ermine a successful predator, able to take down prey larger than itself and dispatch it with bites to the neck. Male Ermines are almost twice the size of females. Adults breed in summer, and after delayed implantation, four to ten young are born in spring, often in a nest in a rodent burrow. The species tends to take advantage of an abundant food supply by killing whatever and whenever it can and then storing surplus, often in a storeroom formed from a side tunnel of its den. This may lead to mass slaughter when the weasel encounters a bounty of potential prey, as in a hen house.

Long-tailed Weasel

Mustela frenata L 10–17 in (25–43 cm)

The largest weasel in North America, its tail measures about half its total length. Coat color shows regional variation, including a "bridled" variety with white markings on the face.

KEY FACTS

Body is brown to reddish to orange to white; underside is often whitish; tail tip is black.

+ habitat: Woodlands, fields, and meadows, usually near water

+ range: Southern Canada and most of U.S., except desert Southwest

+ food: Rodents, rabbits, birds, eggs, snakes, insects, and carrion

Long-tails are a generalized weasel species, occupying a wide geographic range and habitat distribution, and exploiting a wide variety of food sources. Active day and night, the species feeds voraciously to meet its tremendous energy needs, and is equally adept hunting in trees, on the ground, and underground. This weasel shares part of its range with the Ermine and Least Weasel. The Long-tail tends to go after larger prey animals to lessen competition for resources, but it sometimes includes the other weasels on its menu. Like other weasels, Long-tails in the northern parts of the range molt to white coats for the winter.

Black-footed Ferret

Mustela nigripes L 18–23 in (46–58 cm)

The survival of the endangered Black-footed Ferret is tied to the existence of prairie dogs, whose prairie ecosystem has dwindled to about 2 percent of its original expanse.

KEY FACTS

Body is tawny brown; face has black mask; feet and tail tip are black.

+ **habitat:** Great Plains grasslands

+ **range:** Formerly central U.S. to south-central Canada; now, small populations formed by reintro-duced captive-bred animals

+ **food:** Prairie dogs, other ground squirrels, and mice

The masked Black-footed Ferret not only dines on the prairie dog, but also appropriates and remodels the rodent's labyrinthine burrows for its own use. Active mainly at night and at dawn, the ferret hunts mostly underground; when it finds a prairie dog, the ferret seizes it by the throat, kills it, and takes it to its own burrow. The ferret was believed to be extinct until a very small population was found in Wyoming in the 1980s. That population dwindled, and the species again was believed extinct in the wild in 1987. Research suggests that captive-breds familiar with prairie dog burrows before release fare better upon reintroduction.

Least Weasel

Mustela nivalis L 5.5–8 in (14–20 cm)

The smallest carnivore, the Least Weasel packs a lot of predatory punch into a small body that may weigh less than 1 ounce (28 g) or top out around 2 ounces (56 g).

KEY FACTS

Body is brown in summer; throat and underside are white; body in winter is white with black-tipped tail.

+ habitat: Meadows, marshes, brushlands, coniferous and mixed forests

+ range: Alaska through Canada and north-central U.S.

+ food: Mice, voles, other rodents, birds, eggs, and insects

A voracious appetite keeps the Least Weasel on overdrive, alternating periods of intense activity with rest around the clock. Its main quarry are voles and mice and, like other weasels, it may store a surplus in a chamber of its den, which—again, weasel-like—it may have taken over from a burrowing rodent. Unlike other weasels, the female Least Weasel does not experience delayed implantation after mating. She can produce two litters of three to ten young each year, which she nurtures in a den nursery that she sometimes lines with fur from the original occupant victim. The species evades pursuers by squeezing into tight spaces.

American Mink

Neovison vison L 18–24 in (46–61 cm)

The lustrous, warm coat of the American Mink helps it retain heat, especially when it swims. Today, mink farms mainly supply pelts fashioned into coats and other often controversial apparel.

KEY FACTS

Body is brown, often with white patches on chin, chest, and belly; tail tip is black.

+ habitat: Wooded areas, usually near sources of water

+ range: Much of U.S. and Canada, except Southwest and parts of southeastern U.S.

+ food: Small mammals, birds, eggs, fish, crayfish, frogs, and snakes

The voracious, solitary American Mink spends much of its time—by night and at times by day—filling its demanding stomach. It hunts in the water and on land, swimming well with semi-webbed feet and climbing agilely with its streamlined body; it can dive to depths of 20 feet (6 m). The mink's amphibious lifestyle helps it nab a wide variety of prey, including fish and crayfish. Minks mate in the spring; an average of four young with silvery white coats are born in early summer in burrows often appropriated from other mammals, such as Muskrats. When stressed, the species may squeal, hiss, and empty its anal glands.

||||||||||||||||||||||||||||||||||||||||||||||||||||||||||||||||||||||||||||||

American Badger

Taxidea taxus L 24–32 in (61–81 cm)

Like a criminal on the lam, an American Badger in summer often doesn't sleep in the same burrow for two consecutive days. It usually holes up for a longer spell in winter.

KEY FACTS

Body is grayish above; underside is yellowish; darker head has white cheeks and a stripe that extends onto the back.

+ **habitat:** Grasslands, meadows, woodland edges, and deserts

+ **range:** Southwestern Canada to western and central U.S.

+ **food:** Rodents, snakes, birds, invertebrates, and carrion

A low-slung, flat body and long, strong claws make the American Badger a digging machine. It excavates burrows announced at the surface by a hole about a foot wide with a mound of dirt in front. When cornered, a badger may retreat into its burrow or hastily dig a new one, flinging dirt at its pursuer. Badgers are canny hunters, often lying in wait in their prey's empty burrow for a meal delivery. The species is mostly nocturnal and solitary, except during mating season. Females bear one to five cubs in the nest area of a burrow; the cubs stay with her about a month before venturing out and learning to excavate.

American Hog-nosed Skunk

Conepatus leuconotus L 20–35 in (51–89 cm)

A bare, flexible piglike snout allows the American Hog-nosed Skunk to root for insects, grubs, and worms. Its diggings also resemble the work of hogs.

KEY FACTS

Black body has wide white stripe from head to rump; tail is all white or white on top.

+ **habitat:** Desert valleys, brushy canyons, and agricultural areas

+ **range:** Southern Arizona, New Mexico, and Texas, possibly north to Colorado

+ **food:** Insects, worms, rodents, reptiles, and vegetation

Weighing up to 10 pounds (4.5 kg), this skunk's large size and sturdy claws allow it to make quite a disturbance of soil. In Texas this behavior gives it the popular name "rooter skunk." The Hog-nosed Skunk is seldom distracted by the presence of other species during its nocturnal foragings. It usually keeps its head down and continues to root with its wriggling, naked snout. The Hognose has never been a favorite of trappers because its fur is coarse and somewhat dingy. Even rattlesnakes usually retreat rather than face a pungent barrage. The genus name means "little fox"; it is the only skunk genus known in South America.

Hooded Skunk
Mephitis macroura L 24–30 in (61–76 cm)

A ruff of longer neck hair on this long-tailed skunk gives it its common name. The hood can be black or white; in one pattern, the white of the hood may extend down the skunk's back.

KEY FACTS

Body is black with various white patterns on head and back ranging from white hood to side stripes to wide white central stripe.

+ habitat: Streamside brushlands, canyons, grasslands, and deserts

+ range: Southern Arizona and New Mexico; western Texas

+ food: Insects, rodents, bird eggs, and fruit

Despite varied body markings, the Hooded Skunk often has a characteristic thin white line down the center of its face. It lives a life of nocturnal foraging in the desert Southwest, mainly looking for insects, which form the bulk of its diet. Though mostly solitary, several Hoodeds may come together to eat without incident. The species dens often in rocky crevices, but seldom appropriates abandoned cabins or lives under sheds, unlike the Striped Skunk. A female gives birth to one litter of three to five young each year. Like other skunks, Hoodeds have few predators, although they may fall prey to a Coyote or Bobcat.

Striped Skunk

Mephitis mephitis L 21–34 in (53–86 cm)

An encounter with a Striped Skunk in full defensive posture—spine arched, tail lifted—often leads to a spray of stinky fluid that can travel 15 feet (4.6 m) and be smelled a half mile (0.8 km) away.

KEY FACTS

Black body has forked white stripe extending to rump; face has thin, white central stripe; tail is bushy.

+ **habitat:** Fields, open woodlands, and suburbs

+ **range:** Much of U.S. and Canada, except far northern regions

+ **food:** Vegetation, insects, mice, shrews, bird eggs, and carrion

The Striped Skunk and its close relatives use two scent glands at the base of the tail, with caustic, smelly contents, as an extremely successful defense mechanism. The glands contain about three teaspoonfuls of musk that lasts through five or so discharges. Striped Skunks usually hiss and stamp their feet in warning before a discharge, which they seldom have to deploy; most potential predators learn to give the skunk a wide berth. Basically an easygoing species, the skunk adapts well to human habitation. It is sometimes possible to see a mother skunk out on an evening forage with her four to seven kits following in single file.

Western Spotted Skunk

Spilogale gracilis L 12–17 in (30–43 cm)

North America's smallest skunk is known for the distinctive handstand posture with waving tail that it takes when threatening to spray its oily, stinky musk.

KEY FACTS

Body is black with broken white stripes; a white triangle between eyes and a stripe behind eye; tail tip is white.

+ habitat: Deserts, canyons, woodlands, and farmland

+ range: Western U.S. into southern British Columbia

+ food: Rodents, insects, snakes, lizards, bird eggs, and fruit

This agile skunk sports a silky coat marked with a pattern of broken stripes that resemble spots. The species is an adept tree climber and sometimes feeds—and sleeps—in trees. More often it hunts nocturnally on the ground and dens underground or in rock crevices or hollow trees. It may dig its own burrow or take over one abandoned by another animal. The similar Eastern Spotted Skunk *(S. putorius)* is often a bit bigger, its facial triangle is smaller, and the line behind the eye that runs to the mid-back is narrower. It lives in the midwestern United States, extending slightly into Canada, and in parts of the Southeast.

Ringtail

Bassariscus astutus L 26–31 in (66–79 cm)

Also known as miner's cat, the Ringtail was welcome in western gold mining camps, where it ably kept mice populations in check and was tamed to become a companion to prospectors.

KEY FACTS

Body is brownish gray with very long black-banded tail; eyes are large and ringed in white.

+ habitat: Canyons and other rocky areas, mountains

+ range: Throughout much of southwestern U.S.; range is expanding

+ food: Plants, insects, spiders, and rodents, including ground squirrels

The Ringtail is a late-night creature: It tends to come out to hunt among the rocks and scrub after raccoons and skunks have had their fill. It descends from tree perches and canyon walls with great agility, balanced by a tail that measures as long as its body. This super-agile species also has a hind foot that swivels 180°, aiding climbing and making the contents of high-up bird nests easily accessible to a raiding Ringtail. These animals usually are shy and solitary, although sometimes mated pairs may stay together. A litter of one to four offspring are born in the spring in a rock crevice den, burrow, or tree hollow.

White-nosed Coati
Nasua narica L 3.5–4.5 ft (1–1.4 m)

The White-nosed Coati's long, flexible, and sensitive snout allows it to root successfully for food and gives the southwestern species one of its nicknames: hog-nosed coon.

KEY FACTS

Fur is brownish or reddish; face has dusky mask and white spots around the eyes; tail is long and faintly banded.

+ **habitat:** Woodlands, scrublands, canyons, and mountains

+ **range:** Extreme southwestern United States

+ **food:** Insects and other invertebrates, eggs, small vertebrates, and fruit

Unlike its Raccoon and Ringtail cousins, the White-nosed Coati is most active during the day. It lives in bands of 30 or more females and young, sleeping in trees and coming down at dawn to forage. Adult males are usually solitary (another nickname is *gato solo*—"lone cat"), and are welcomed by females only during mating season. Females give birth to litters of two or more in early summer. A coati walks with its long tail upright; the tail helps it balance when climbing trees, but is not prehensile. Primarily a Central American species, the White-nosed Coati ranges into southern Texas, New Mexico, and Arizona.

Raccoon

Procyon lotor L 22-39 in (56-99 cm)

Humanlike hands with thin, dexterous fingers and a delicate sense
of touch give the Raccoon a great advantage when searching for
food on land, in the water—or in your latched garbage can.

KEY FACTS

Body is grizzled
gray; nose and face
mask are black; tail is
striped. Color varies
regionally.

+ habitat: Woods,
wetlands, suburban
and urban areas

+ range: Most of U.S.
(except parts of West)
and southern Canada

+ food: Berries, nuts,
seeds, crayfish, crabs,
fish, turtles, eggs,
small mammals, human
food

The Raccoon has a large skill set that allows it to adapt
easily to urbanization. The mostly nocturnal species
typically spends its days in a tree, coming down at night
to pursue an omnivore lifestyle. It often forages for prey
in water, using fingers to manipulate, but not wash its
catch. Raccoons mate in early spring, and by April or May
the two to seven kits are born in a tree
hollow den—or perhaps in your
attic. Raccoons sometimes den
communally and do not hibernate.
They can carry rabies and parasites;
disoriented or aggressive raccoons should
always be avoided. Call local animal control to
report one.

Burro/Donkey
Equus asinus H at withers 4 ft (1.2 m)

Descended from strays or castoffs from the 16th century onward, about 6,000 burros roam the southwestern United States. These hardy animals have no natural predators in their range.

KEY FACTS

Shaggy fur color is variable, ranging from red to brown to gray; mane is erect; ears are long; face often has white snout and eye rings; tail is tufted.

+ habitat: Canyons, mountains, and deserts

+ range: Western U.S., especially in California, Nevada, Arizona, and New Mexico

+ food: Grasses, shrubs, and cacti

The Burro served as a sure-footed beast of burden in the settlement of the West. Left to fend for itself in the wild, this close relative of the horse had the necessary skills to survive. In deserts, where browse and water are at a premium, it competes with native species—a fact that led in the past to elimination and more recently to relocation from public lands such as the Grand Canyon. The female, or jenny, bears a foal after a gestation of about 12 months. Jennies and young form stable groups. Males, or jacks, may dominate a territory or wander with other bachelors, braying and battling for mating opportunities.

Horse

Equus caballus H at withers 5 ft (1.5 m)

North America's native horse disappeared at least 8,000 years ago, perhaps a result of Stone Age hunting. Wild populations now consist of feral horses descended from imported breeds.

KEY FACTS

Varying colors of brown, bay, black, white, palomino, or pinto are common; animals tend to be small in height and weight.

+ **habitat:** Range-lands and islands

+ **range:** British Columbia and Nova Scotia; western U.S.; southeastern U.S. barrier islands

+ **food:** Grasses and forbs

Feral Horses derive mainly from Spanish imports, U.S. Cavalry retirees, and workhorses. Most herds occur in western states and British Columbia. Three to 20 animals or more usually are headed by a dominant stallion that controls his harem of females and often an older mare. After an 11-month gestation, mares give birth to one colt, which stays with the herd until age three or four. Assateague Island, a barrier island divided between Maryland and Virginia, is home to two populations of wild "ponies" that have sorted themselves into small bands. How to respond to overpopulation of western herds is a controversial issue.

Wild Boar

Sus scrofa H at shoulder 24–42 in (61–107 cm)

This introduced Old World member of the swine family has long interbred with the feral domestic pig, but as a wild species it spends its life mostly unimpeded in much of its range.

KEY FACTS

Body is coarse, brownish or blackish; tusks curve upward; young have stripes and spots.

+ habitat: Marshes, forests, brushlands, and mountains

+ range: New Hampshire and scattered in U.S. South and West along with feral pigs and hybrids

+ food: Roots, tubers, leaves, nuts, fruit, fungi, insects, invertebrates, vertebrates, and eggs

Wild Boars leave many marks on the land, ranging from mud wallows, to extensive swaths of bulldozed earth, to scraped-up trees. They often destroy the habitats of native plants and animals in the process. The boars forage at night, dawn, and dusk, or in the daytime in winter. They are formidable foes, slashing with razor-sharp tusks and sharp hooves. The females and young travel in small bands; males are mostly solitary. Females often bear two litters of 3 to 12 young. Wild Boars rest often, in mud wallows or under a shady canopy they construct by incorporating cut grass with standing vegetation.

Collared Peccary/Javelina

Tayassu tajacu or *Pecari tajacu* H at shoulder 15–20 in (38–51 cm)

These piglike hoofed mammals belong to a group that diverged from a common ancestor more than 40 million years ago.

KEY FACTS

Body is grizzled with pale yellow collar; long snout; downward-pointing tusks.

+ **habitat:** Deserts, canyons, and mixed forest

+ **range:** Southern Arizona, southwestern New Mexico, and southern Texas

+ **food:** Cacti, shrubs, fruit, and nuts

Known as the Javelina for its razor-sharp tusks, the Collared Peccary has a complex stomach well suited to the challenge of one of its favorite foods—Prickly Pear Cactus, which it eats spines and all. Peccaries live in mixed-sex social groups of 5 to 15 individuals. They communicate by means of glands on their rumps that exude an oily, musky fluid. They mark each other with the scent and rub against rocks and trees to delineate a home range. They observe no particular mating season, but males will posture and fight with head-on charges in which their tusked jaws occasionally lock. Twins are usually born four months after mating.

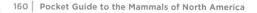

Moose

Alces alces H at shoulder to 7.5 ft (2.3 m)

The largest deer species, the solitary behemoth of the northern woods has a large "Roman" nose and ungainly, high-shouldered appearance. A bull Moose can weigh up to 1,800 pounds (800 kg).

KEY FACTS

Body is blackish brown; legs are usually paler; young are reddish brown.

+ habitat: Northern forests, tundra, willow thickets, and swamps

+ range: Alaska through Canada and into Rockies, north-central U.S. and New England

+ food: Leaves, twigs, bark of trees, aquatic plants

Moose browse on plant parts—the name comes from the Algonquian for "he cuts or trims smooth"—and they also wade out into ponds in summer to feed on pondweed and water lilies. A bull Moose's immense rack of palmate antlers, part of his mating-time appeal, may weigh 70 pounds (32 kg). Rival bulls engage in headfirst shoving matches, but the cows may also lure males with grunting, mooing calls, which hunters often imitate. Cows give birth to one calf, but sometimes twins, after an eight-month gestation. Unlike other young deer, a Moose calf lacks spots; its formidable, aggressive mother can keep a wolf at bay, if need be.

Mule Deer

Odocoileus hemiones H at shoulder 3–3.5 ft (0.9–1 m)

Ears that are two-thirds the length of its head give the Mule Deer its name. The large appendages help the western species detect danger at long range.

KEY FACTS

Body is gray to brown to red: rump patches are whitish; slim tail has a black tip.

+ **habitat:** Deserts, brushlands, coniferous and mixed forests, and mountains

+ **range:** Southwest Alaska and southwestern Canada through western U.S.

+ **food:** Twigs, barks, leaves, buds, nuts, and berries

Well adapted to high elevations, broken country, and arid regions, Mule Deer often cover rugged landscapes. They are also known as "jumping deer" because of a method of locomotion called stotting. It consists of bounding stiff-legged with all four feet off the ground at the same time. Adult females and their offspring form groups; bucks usually are solitary. Mating occurs in the fall, and does give birth to one young or twins, depending on their age. Subspecies include the Columbia Black-tailed Deer (*O. h. columbianus*) and the Sitka Deer (*O. h. sitkensis*), which inhabit the Pacific Northwest and British Columbia.

White-tailed Deer

Odocoileus virginianus H at shoulder 2–3.5 ft (0.6–1 m)

In some parts of the continent, White-tailed Deer seem to be becoming as common as Gray Squirrels. The adaptable deer species is pushing ever farther into urban environments.

KEY FACTS

Fur is grayish brown in summer, reddish in winter; underside is white.

+ **habitat:** Deciduous woodlands, brush-lands, and scrublands

+ **range:** U.S. and Canada south of Hudson Bay, except parts of southwestern U.S.

+ **food:** Leaves, twigs, nuts, berries, and fungi; crops and ornamental plants

White-tailed bucks grow antlers in the fall that are covered in a blood-rich "velvet" that they scrape off against trees; the antlers fall off after the rut. The species lives in groups of females and young and bachelor males. In winter, larger, mixed groups form in forage areas. Does have a single fawn at first, followed by twins in later years; the spotted young are left in cover while the mother feeds. In many areas, herds are culled regularly to keep populations in check. The tiny Key Deer (*O. virginianus clavium*), an endangered subspecies found only in the Florida Keys, was isolated there at the end of the last ice age.

Caribou

Rangifer tarandus H at shoulder 2.3–4.5 ft (0.7–1.4 m)

Twice a year, tens of thousands of Caribou migrate 900 miles (1,400 km) across the tundra. They head to calving grounds in the spring and return to winter pastures in the fall.

KEY FACTS

Head, body, and legs are brown, neck and rump are pale; male's antlers are large and elaborate, female's are smaller, thinner.

+ habitat: Boreal coniferous forest and tundra

+ range: Alaska through Canada to extreme northeast Washington and western Idaho

+ food: Grasses, sedges, forbs, shrubs, and lichens

When the Caribou isn't migrating, it travels more locally in search of forage in small bands or loose herds of up to several hundred animals. The bulk of its diet, especially in winter, consists of lichens, which it nibbles with small, weak teeth. Adult Caribou eat about 12 pounds (5.5 kg) of the slow-growing lichen each day. Subspecies of Caribou in North America include *R. t. groenlandicus* in Canada and *R. t. granti* in Alaska, known as "barren-ground" Caribou. In spring, Caribou cows in a herd drop their single calves in a synchronized fashion within days of each other, and the precocial young are soon on the move.

Elk/Wapiti

Cervus elaphus H at shoulder up to 5 ft (1.5 m)

Shawnee Indians called this highly evolved deer "Wapiti." Largely
a western species now, the Elk is adapted to grazing and browsing.
At least one herd member keeps watch while the others eat.

KEY FACTS

Body is light brown;
head, neck, and legs
are dark brown; rump
patch and tail are pale;
male grows antlers.

+ habitat: Plains,
meadows, woodlands,
and mountains

+ range: Western
U.S. to southwestern
Canada; introduced
or reintroduced
elsewhere

+ food: Grasses,
forbs, and woody
plants

Like a lot of large mammals in North America, the Elk
once roamed a larger area; it covered much of the
United States and Canada. This stately deer establishes
large herds containing hundreds of animals in open areas
and smaller numbers in woodlands. Bull Elk join the cows
and young during the fall rut when they spar with other
males for mating opportunities. They posture and challenge rivals, and when they
meet head-on, their antlers—with their
single main branches—seldom lock.
Females give birth to a single calf in
the late spring. The Tule Elk is a small
and pale subspecies that lives in east-
central California.

Pronghorn

Antilocapra americana H at shoulder 2.5–3.5 ft (0.8–1 m)

Antilocapra means "antelope goat," but the Pronghorn does not represent either family. It is a remnant of a group of horned mammals that arose in North America some 50 million years ago.

KEY FACTS

Tan with white bands on cheeks, sides, and underside; black markings on male's head and mane.

+ habitat: Grasslands, sagebrush, and desert

+ range: Scattered populations in southwestern Canada and western U.S.; reintroduced elsewhere

+ food: Grasses, forbs, cacti, and shrubby browse

The fastest Northern American mammal, the Pronghorn can reach speeds of more than 50 miles (80 km) an hour for short spurts. It can raise and fan the white hairs of its rump patch as a warning to others. The species has a cruising speed of about 25 miles (40 km) an hour. Adult males sport forked horns, not antlers, and shed only the outer sheath; females have a prongless version shorter than their ears or no horns at all. Dominant males control a territory and the females within it. Usually twins are born to the doe in the spring, and within days they can run faster than a human. Pronghorns once roamed the plains in the millions.

||||||||||||||||||||||||||||||||||||||||||||||||||||||||||||||||||||||||||||

American Bison/Buffalo

Bos bison or *Bison bison* H at shoulder up to 6 ft (1.8 m)

This iconic, cowlike behemoth of the Great Plains was slaughtered almost to extinction by 1900. Today, American Bison live mostly in protected herds.

KEY FACTS

Adult is dark brown, curly on head; calf is reddish; both sexes have short, curved horns.

+ habitat: Plains, prairies, and woodlands

+ range: Small, very scattered populations from Alaska through western Canada and western U.S.

+ food: Grasses and browse

The American Bison, erroneously known as the Buffalo, once roamed the plains in vast herds. Native peoples hunted them and used every part for various needs. The bison is front-loaded with enormous shoulders and a massive head with pointed horns. Bulls weighing about a ton (900 kg) join herds of females and subadults during the rut and may lose 300 pounds (130 kg) by its end. Calves are born nine months later and are mobile after a few hours. The Wood Bison (*B. bison athabascae*) of Canada is larger and darker than the plains species, and it lives in protected herds that are kept separate to preserve genetic differences.

Mountain Goat

Oreamnos americanus H at shoulder 3–4 ft (0.9–1.2 m)

Casually jumping from ledge to ledge on rock faces, steadied by flexible, rubbery, shock-absorbing hooves, the Mountain Goat owns the high crags of northwestern North America.

KEY FACTS

Body is white; eyes, nose, horns, and hooves are black; males, females, and young have stiletto horns.

+ habitat: Steep mountain areas

+ range: Southern Alaska and western Canada and to Washington, Wyoming, and Idaho

+ food: Alpine plants and shrubs

In its rocky realm, the Mountain Goat is mostly safe from predators, apart from an occasional airborne eagle snatching a kid. The landscape is a bigger threat: Avalanches and rockslides take a toll on the species. Forays into mountain meadows also make it vulnerable. The adult female Mountain Goat is a formidable foe, fending off any threat to her kids. A male often takes a juvenile approach to courting, crawling on his belly and squeaking like a baby to win a nanny's favor. After mating, he retreats. One to three kids are born after six months; they're immediately agile and rock worthy, and stay with their mother for about a year.

Muskox

Ovibus moschatus H at shoulder 3–5 ft (0.9–1.5 m)

The Muskox's scientific name means "musky sheep ox," but it's a misnomer. Muskoxen are neither sheep nor oxen, and they lack musk glands, although bull urine is pungent.

KEY FACTS

Body is shaggy dark brown; shoulders and rump are humped; both sexes have large, curved horns.

+ **habitat:** Arctic tundra

+ **range:** Arctic islands and coastal plains of Northwest Territories in Canada; reintroduced elsewhere

+ **food:** Grasses, woody plants, willows, mosses, and lichensa

When threatened, Muskox herds form a defensive ring, with their formidable heads facing out and the calves secured in the center behind a rampart of adult bulk. If the threat escalates, an adult will rush to gore an intruder and if successful, the rest may trample it. This tactic works well with wolves, but has made the animals an easy target for human hunters from prehistoric times.

Muskoxen graze on the tundra, moving between summer and winter ranges. Their underwool is featherlight and incredibly warm. The Inuit wove mosquito nets from the long guard hairs, which can measure 24 inches (61 cm) on an adult male.

Bighorn Sheep

Ovis canadensis H at shoulder 3.5 ft (1 m)

The Bighorn Sheep lacks the pinpoint agility of the Mountain Goat, but it still takes command of rock ledges in western mountains. A 2-inch (5 cm) path is all it needs for travel.

KEY FACTS

Body is tan to dark brown; male has massive, tightly curled horns; female is smaller with narrower and straighter horns.

+ habitat: Mountain areas

+ range: Rocky Mountains from Canada south to New Mexico and western Texas

+ food: Grasses, sedges, and woody plants

Few animal encounters are as dramatic as the challenge of two Bighorn bucks for dominance. Posturing leads to the head-on crash of two heavy-horned and reinforced skulls, a sound that can be heard from long distances. They continue until one concedes—or dies. With victory comes the opportunity to mate. The males' horns can weigh 30 pounds (14 kg) and show annual growth rings. These sheep travel in herds; females and young stay together as do bachelor males. Desert Bighorns of the southern Rockies and western Texas are smaller and leaner. They obtain moisture from succulent plants and are well adapted to arid lands.

|||||||||||||||||||||||||||||||||||||||||||||||||||||||||||||||||||||||||||||||||||

Dall's Sheep

Ovis dalli H at shoulder 3.5 ft (1 m)

Dall's Sheep living in mountain terrain are wary of predators, but in protected areas, they show an inquisitiveness that may have led to the early domestication of other sheep.

KEY FACTS

Body is pure white; male has large, darker, spiraled horns; female horns are thinner and straighter.

+ habitat: Mountainous regions

+ range: Alaska, Yukon, western Northwest Territories, and northern British Columbia

+ food: Grasses, sedges, forbs, and woody plants

Dall's Sheep roam in bands separated by sex: Bachelor groups composed of old and young males travel separately from females and juveniles. Young males appease the established adults by behaving like females. The sexes come together only during the mating season. Dall's Sheep move between grazing patches that may be 40 miles (64 km) apart. Young animals learn the winding routes from their elders. The species belongs to the "thinhorn" branch of the genus *Ovis*. Their horns spiral widely away from the head. Stone Sheep (*O. d. stonei*) is a darker subspecies that overlaps with Dall's in the southern part of the range.

Natural Regions of North America

The continental United States and Canada can be divided into nine physiographic regions.

Appalachian Highlands

The oldest mountain chain in North America, the heavily eroded Appalachians extend for about 2,000 miles (3,200 km) from Alabama to Newfoundland, with the highest peak being Mount Mitchell, in North Carolina, at 6,684 feet (2,040 m). This mountain range started forming 480 million years ago, when continental collisions caused volcanic activity and mountain building. Ecologically, this region hosts eastern temperate forests with a great variety of coniferous and deciduous trees and wildflowers, some of them high-elevation specialists.

Coastal Plain

This gradually rising flatland spans some 4,000 miles (6,400 km) in total and covers several distantly related regions along the Gulf of Mexico, the southern Atlantic coast, and the northernmost coasts of the Arctic Ocean. In the past, oceans covered these plains, depositing sediment layers over millions of years, until falling sea levels exposed them. The types of vegetation in these widely separated regions range from subtropical trees and flowers in the southernmost Coastal Plain to marsh plants along the mid-Atlantic to the largely treeless tundra on the northern coast of Alaska.

Interior Plains

Ranging from the lowlands of the St. Lawrence River Valley in the east to the mile-high Great Plains in the west, the vast Interior Plains of North America provide fertile soils, especially for productive prairie farms. A shallow sea covered much of this region as recently as 75 million years ago, and sediments from rivers draining the Appalachians and western mountains were deposited in layers throughout the sea. The Great Plains were once covered by vast and diverse expanses of natural grasses, sagebrush, and a varied suite of wildflowers. Much of this ecosystem has vanished, the land brought into use by modern agriculture and extensive grazing.

Interior Highlands

The Ozark Plateau and Ouachita Mountains form the Interior Highlands, which are centered on Arkansas and southern Missouri, with mountains reaching more than 2,600 feet (800 m) high. These ancient eroded highlands were connected to the Appalachians until tectonic activity separated them some 200 million years ago. Ecologically, this relatively small area straddles the southern Interior Plains and the eastern Coastal Plain, with trees and wildflowers representing both regions.

Rocky Mountains

The highest mountain system in North America, the Rockies dominate the landscape for some 3,000 miles (4,800 km), from New Mexico
to Alaska, with more than 50 peaks surpassing 14,000 feet (4,300 m).
Tectonic activity uplifted the Rockies about 50 to 100 million years
ago, making them much younger and less eroded than the Appalachians. The ecological hallmarks of the Rockies are its coniferous
forests of pines, firs, and spruces, adapted to high elevations, with
wildflower species similarly adapted to elevations and temperatures.

Intermontane Basins and Plateaus

This region is called intermontane because it is situated between
the Pacific and the Rocky Mountain systems. Pacific mountains block
most moisture-bearing clouds coming from the Pacific Ocean, giving desert climates to places like the Colorado Plateau, 5,000 to
7,000 feet (1,500 to 2,100 m) high, and the Great Basin. In Canada
and Alaska, the immense Yukon River Valley and the Yukon-Tanana
Uplands are part of this region. The deserts feature cacti and a host
of other specialist plants of the arid West. Cottonwoods, ashes, and
willows line rivers that run intermittently through the dry plains.

Pacific Mountain System

From Alaska to California, mountains and volcanoes tower over the
western coast in an almost unbroken chain. These mountain ranges
are geologically young and seismically active, with uplift starting
some five million years ago. The highest mountain in North America,
Alaska's Mount McKinley (20,320 feet/6,200 m), is still growing at about
a millimeter a year—about the thickness of a fingernail. Distantly separated from the Rockies, this mountain range supports its own distinct
varieties of trees and wildflowers adapted to higher elevations.

Canadian Shield

The geologic core of North America is the Canadian Shield, which
contains the continent's oldest rocks. Landforms are relatively flat,
having been eroded and scoured by glaciers over millions of years.
The exposed bedrock ranges in age from 570 million to more than

3 billion years old. This is a vast and extensively diverse region of climatic extremes and varied vegetation, from dense boreal forests in the south to frigid tundra in the north, populated by stunted trees, small shrubs, lichens, and ground-clinging herbs.

Arctic Lands

Highlands known as the Innuitian Mountains cover most islands. The icy climate is too harsh for most animals and vegetation, and much of the ground is permanently frozen. Nevertheless, low-growing shrubs, small tundra plants, and lichens manage to survive.

Further Resources

BOOKS

Elbroch, Mark, and Kurt Rinehart. B*ehavior of North American Mammals*. Peterson Reference Guides. Houghton Mifflin, 2011.

Mammal Species of the World: A Taxonomic and Geographic Reference, 3rd ed. Don E. Wilson and DeeAnn M. Reeder (editors). Johns Hopkins University Press, 2005.

National Geographic Book of Mammals. National Geographic Society, 1998.

Reid, Fiona A. *A Field Guide to the Mammals of North America*, 4th ed. Peterson Field Guides. Houghton Mifflin, 2006.

Rezendes, Paul. *Tracking and the Art of Seeing: How to Read Animal Tracks and Signs*, 2nd ed. Harper Collins, 1999.

WEBSITES

eNature.com

National Geographic Society: animals.nationalgeographic.com/animals/mammals

Smithsonian Institution: www.mnh.si.edu/mna

APPS

Audubon Mammals: A Field Guide to North American Mammals. Green Mountain Digital.

About the Author

 **CATHERINE HERBERT HOWELL,** a former National Geographic staff member, has written extensively on nature and natural history. She explored the relationships between people and plants in *Flora Mirabilis: How Plants Have Shaped World Knowledge, Health, Wealth, and Beauty* (2009) and covered the importance of birds in world cultures in *National Geographic Bird-watcher's Bible* (2012). Howell also wrote *National Geographic Pocket Guide to Wildflowers of North America* (2014) and *National Geographic Pocket Guide to Reptiles and Amphibians of North America* (2015) and was a contributing writer to *National Geographic Illustrated Guide to Nature* (2013) and *National Geographic Illustrated Guide to Wildlife* (2014). She serves as a master naturalist volunteer in Arlington, Virginia.

About the Artist

JARED TRAVNICEK is a medical illustrator with an M.A. in medical and biological illustration from the Johns Hopkins University School of Medicine. He is based in Indianapolis, Indiana, where he works as a neurosurgical illustrator.

Acknowledgments

I would like to thank the National Geographic editorial team, especially Susan Tyler Hitchcock, for the opportunity to work on this book. Thank you to Barbara Payne, Zachary Galasi, Paul Hess, Noelle Weber, Uliana Bazar, Patrick Bagley, and the others on the National Geographic team. Thanks also to zoologist Sam Zeveloff of Weber State University for his expert guidance. As always, my colleagues in the Arlington Regional Master Naturalists and the naturalists at Long Branch Nature Center continue to inspire my efforts.

Illustrations Credits

Front Cover: (wolf), Norbert Rosing/National Geographic Creative; (chipmunk), npine/ Shutterstock; (bobcat), Altrendo Nature/Getty Images; (fox), Michael DeYoung/Design Pics/National Geographic Creative; (mountain goat), Josh Schutz/Shutterstock.

Spine: Heiko Kiera/Shutterstock.

Back Cover: (arctic hare), Jim Brandenburg/Minden Pictures/National Geographic Creative; (bat), Merlin Tuttle/BCI/Getty Images; (armadillo), Stan Tekiela/NatureSmart Wildlife; (grizzly), Paul Souders/Getty Images.

2-3, Jason Savage/TandemStock.com; 4, Loic Poidevin/naturepl.com; 7, Joel Sartore/ Getty Images; 9, J. L. "Woody" Wooden/Getty Images; 12, David Courtenay/Getty Images; 13, Stan Tekiela/NatureSmart Wildlife; 14, Wayne Lynch/Getty Images; 15, Visuals Unlimited, Inc./Don Grall/Getty Images; 16, Michael G. Mill/Shutterstock; 17, Thomas & Pat Leeson/Science Source; 18, iStock.com/mandj98; 19, PHOTO 24/Getty Images; 20, Stan Tekiela/NatureSmart Wildlife; 21, Ed Cesar/Science Source; 22, Nicholas Bergkessel, Jr./Science Source; 23, Stubblefield Photography/Shutterstock; 24, Martha Marks/Shutterstock; 25, Robert Franz/Getty Images; 26, Henk Bentlage/Shutterstock; 27, Fremme/Shutterstock; 28, Craig K Lorenz/Getty Images; 29, Andreas Resch/ Shutterstock; 30, Tom Reichner/Shutterstock; 31, Stan Tekiela/NatureSmart Wildlife; 32, Ed Reschke/Getty Images; 33, worldswildlifewonders/Shutterstock; 34, Kenneth W. Fink/Getty Images; 35, Sam Fried/Science Source; 36, Stan Tekiela/NatureSmart Wildlife; 37, Tim Stirling/Shutterstock; 38, iStock.com/brokentone; 39, Stan Tekiela/ NatureSmart Wildlife; 40, Howard Stapleton/Alamy; 41, Tom Reichner/Shutterstock; 42, Francis Bossé/Shutterstock; 43, Danita Delimont/Getty Images; 44, iStock.com/Charles Schug; 45, Purestock/Getty Images; 46, Yva Momatiuk & John Eastcott/Minden Pictures/ National Geographic Creative; 47, C. Allan Morgan/Getty Images; 48, Stan Tekiela/ NatureSmart Wildlife; 49, Shattil & Rozinski/NPL/Minden Pictures; 50, John Cancalosi/ Getty Images; 51, E.R. Degginger/Alamy; 52, Gary Meszaros/Science Source; 53, Wayne Lynch/Getty Images; 54, Tom McHugh/Getty Images; 55, Daniel Cox/Getty Images; 56, Kenneth L. Crowell/Mammal Images Library; 57, Stan Tekiela/NatureSmart Wildlife; 58, Stan Tekiela/NatureSmart Wildlife; 59, Rob & Ann Simpson/Visuals Unlimited, Inc.; 60, Robert J Erwin/Getty Images; 61, Barry Mansell/Minden Pictures; 62, Brian Lasenby/ Shutterstock; 63, David Moskowitz, davidmoskowitz.net; 65, Stan Tekiela/NatureSmart Wildlife; 66, Rolf Kopfle/Getty Images; 67, Rob & Ann Simpson/Visuals Unlimited, Inc.; 68, Barry Mansell/naturepl.com; 69, Stan Tekiela/NatureSmart Wildlife; 70, Stan Tekiela/ NatureSmart Wildlife; 71, Stan Tekiela/NatureSmart Wildlife; 72, Gary Meszaros/ Visuals Unlimited/Getty Images; 73, Michael Durham/Minden Pictures; 74, Rick & Nora Bowers/Alamy; 75, Louise Murray/age fotostock; 76, Stan Tekiela/NatureSmart Wildlife; 77, Stan Tekiela/NatureSmart Wildlife; 78, Rodger Jackman/Getty Images; 79, Erni/ Shutterstock; 80, Stan Tekiela/NatureSmart Wildlife; 81, Tom Reichner/Shutterstock; 82, Steven Kazlowski/Getty Images; 83, Stan Tekiela/NatureSmart Wildlife; 84, Howard Sandler/Shutterstock; 85, Jim Brandenburg/Minden Pictures/National Geographic Creative; 86, iStock.com/twildlife; 87, David Tipling/Getty Images; 88, Pyshnyy Maxim Vjacheslavovich/Shutterstock; 89, A & J Visage/Alamy; 90, Stan Tekiela/NatureSmart Wildlife; 91, Hal Beral/Visuals Unlimited, Inc.; 92, Tom Reichner/Shutterstock; 93, altrendo nature/Getty Images; 94, visceralimage/Shutterstock; 95, John Macgregor/ Getty Images; 96, Rob & Ann Simpson/Visuals Unlimited, Inc.; 97, Phil Myers; 98, Tom & Pat Leeson; 99, Stan Tekiela/NatureSmart Wildlife; 101, Gary Meszaros/Visuals Unlimited, Inc.; 103, Ronald G. Altig; 104, Dwight R. Kuhn; 105, Ken Catania/Visuals Unlimited, Inc.; 106, Dwight R. Kuhn; 107, Gerry Ellis/Minden Pictures/National Geographic Creative; 108, Michael Durham/www.DurmPhoto.com; 109, Merlin Tuttle/BCI/Getty Images; 110, Merlin Tuttle/Getty Images; 111, Pete Oxford/Getty Images; 112, Joel Sartore/Getty Images; 113, Jared Hobbs/Getty Images; 114, Merlin D. Tuttle/Science Source; 115,

Index

Boldface indicates
species profile.

A

Alces alces **161**
Ammospermophilus
23, 24
Antelope Squirrel
Harris's **23**
White-tailed **24**
Antilocapra americana
166
Antrozous pallidus
121
Aplodontia rufa **14**
Armadillo, Nine-
banded/Common
Long-nosed **13**

B

Badger, American **149**
Bassariscus astutus **154**
Bat
Big Brown 11, **113**
Brazilian Free-tailed/
Mexican Free-tailed
112
California Leaf-nosed
110
Eastern Pipistrelle **119**
Eastern Red **114**, 117
Hoary **115**
Little Brown/Little
Brown Myotis **124**
Long-eared Myotis
123
Northern Yellow **116**
Pallid **121**
Rafinesque's Big-eared
120
Seminole/Mahogany
116, **117**
Silver-haired **122**

Southern Long-nosed/
Lesser Long-nosed
109
Southern Yellow 116
Western Mastiff **111**
Western Pipistrelle
118
Bear
American Black **136**
Grizzly/North
American Brown
11, **137**, 138
Kodiak **138**
Polar 133, **139**
Beaver, American 11,
44, 62, 140
Bison
American 10, **167**
Wood 167
Bison bison **167**
Blarina brevicauda **95**
Boar, Wild **159**
Bobcat 12, 25, 61, 87,
126, **127**, 151
Bog Lemming
Northern 54, 65
Southern **65**
Bos bison **167**
Brachylagus idahoensis
82
Buffalo **167**
Burro **157**

C

Callospermophilus
lateralis **37**
Canis **129, 130, 131**
Caribou **164**
Castor canadensis **44**
Catamount. see
Mountain Lion
Cervus elaphus **165**
Chaetodipus
californicus **50**
Chipmunk
Alpine **38**

Cliff **39**
Eastern **42**
Least 38, **41**
Merriam's **40**
Townsend's **43**
Western 37
Clethrionomys gapperi
60
Coati, White-nosed **155**
Collared Lemming
Northern **53**, 54
Ungava **54**
Condylura cristata **105**
Conepatus leuconotus
150
Cony **81**
Corynorhinus
rafinesquii **120**
Cotton Rat, Hispid **77**
Cottontail
Desert **90**
Eastern 89, 90, **92**
Mountain **93**
New England 92
Cougar. see Mountain
lion
Coyote 10, 12, 38, 39,
44, 74, **129**, 131, 151
Coypu **46**
Cryptotis parva **96**
Cynomys **25, 26**

D

Dasyapus novemcintus
13
Deer
Columbia Black-tailed
162
Key 163
Mule **162**
Sitka 162
White-tailed **163**
Deermouse
Brush **71**, 74
North American 72,
73

National Geographic
Pocket Guide to the Mammals
of North America

Catherine Herbert Howell

Published by the National Geographic Society
Gary E. Knell, *President and Chief Executive Officer*
John M. Fahey, *Chairman of the Board*
Declan Moore, *Chief Media Officer*
Chris Johns, *Chief Content Officer*

Prepared by the Book Division
Hector Sierra, *Senior Vice President and General Manager*
Lisa Thomas, *Senior Vice President and Editorial Director*
Jonathan Halling, *Creative Director*
Marianne R. Koszorus, *Design Director*
Susan Hitchcock, *Senior Editor*
R. Gary Colbert, *Production Director*
Jennifer A. Thornton, *Director of Managing Editorial*
Susan S. Blair, *Director of Photography*
Meredith C. Wilcox, *Director, Administration and Rights Clearance*

Staff for This Book
Barbara Payne, *Editor*
Paul Hess, *Text Editor*
Zachary Galasi, *Project Editor*
Sanaa Akkach, *Art Director*
Uliana Bazar, Patrick J. Bagley, *Photo Editors*
Noelle Weber, *Designer*
Carl Mehler, *Director of Maps*
Marshall Kiker, *Associate Managing Editor*
Judith Klein, *Senior Production Editor*
Rock Wheeler, *Rights Clearance Specialist*
Katie Olsen, *Design Production Specialist*
Nicole Miller, *Design Production Assistant*
Rachel Faulise, *Manager, Production Services*

Your purchase supports our nonprofit work and makes you part of our global community. Thank you for sharing our belief in the power of science, exploration, and storytelling to change the world. To activate your member benefits, complete your free membership profile at natgeo.com/joinnow.

The National Geographic Society is one of the world's largest nonprofit scientific and educational organizations. Its mission is to inspire people to care about the planet. Founded in 1888, the Society is member supported and offers a community for members to get closer to explorers, connect with other members, and help make a difference. The Society reaches more than 450 million people worldwide each month through *National Geographic* and other magazines; National Geographic Channel; television documentaries; music; radio; films; books; DVDs; maps; exhibitions; live events; school publishing programs; interactive media; and merchandise. National Geographic has funded more than 10,000 scientific research, conservation, and exploration projects and supports an education program promoting geographic literacy. For more information, visit www.nationalgeographic.com.

For more information, please call 1-800-NGS LINE (647-5463) or write to the following address:

National Geographic Society
1145 17th Street N.W.
Washington, D.C. 20036-4688 U.S.A.

For information about special discounts for bulk purchases, please contact National Geographic Books Special Sales: ngspecsales@ngs.org

For rights or permissions inquiries, please contact National Geographic Books Subsidiary Rights: ngbookrights@ngs.org

ISBN: 978-1-4262-1648-0

Printed in China

15/RRDS/1

PACK WHAT MATTERS

And explore with the experts!

Slim • Affordable • Easy-to-use field guides

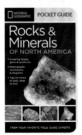